SMALLHOLDING

SMALLHOLDING

Alan Beat

Smallholding Press

First published in the UK in 2015
by Smallholding Press

ISBN 978-0-95-469231-5

Smallholding Press
The Bridge Mill
Bridgerule
Holsworthy
Devon
EX22 7EL

Email: alan@thebridgemill.org.uk

By the same author: *A start in smallholding*
published by Smallholding Press in 2004

Contents

Acknowledgements

Everything in this book is the result of personal experience - mostly ours. For the few exceptions, I have drawn on the first-hand experience of other smallholders. I wish to thank Ritchie and Pammy Riggs, Alison Wilson, Phil Chandler and Penny Hall for generously sharing their specialist knowledge. Thanks also to Liz Conner, and Roger and Alison Parmley, for allowing their livestock to be photographed.

I am grateful to Katie Thear for accepting my early articles for publication in her magazine *Home Farm*; to Simon McEwan, current editor of the magazine, now *Country Smallholding*, for his continuing encouragement; and to David King for commissioning me to write this book (although changed circumstances prevented him from publishing it).

Special thanks to Christine Parsons for agreeing to sell The Bridge Mill to us, and for her unfailing support and friendship ever since.

Most of all, my heartfelt thanks to Rosie, Katie and Martin for sharing the journey into smallholding with me.

Foreword

by Simon McEwan, Editor of *Country Smallholding* magazine

These are exciting times for the self-sufficiency movement. There has been a grow-your-own revolution, with more and more people wanting a taste of "the good life". Sales of vegetable seeds have increased dramatically, there are long waiting lists for allotments, backyard chicken-keeping is booming and other traditional smallholding livestock, such as pigs and bees, are surging in popularity. This interest has also been reflected in a plethora of TV programmes and newspaper articles about self-sufficiency.

The modern movement has its origins in the 1970s, when influential figures such as John Seymour and Katie Thear espoused the benefits of self-sufficiency, and the TV programme *The Good Life* helped to bring it to national awareness. Back then, it was still regarded as rather cranky; now, it is mainstream.

There are several factors fuelling this movement. One is concern about healthy eating alongside the realisation that home-grown food tastes far better than bland supermarket fare. Another concern is for the environment, and for the miles racked up by trucking supermarket food around the country; whereas food from your own garden or allotment travels just a matter of feet. Then there is food security and the realisation that Britain's food supply is really quite tenuous, with much of it imported. It seems so much safer to grow your own.

But the key factor is surely quality of life and a yearning to reconnect with nature. In the soulless world of concrete, plastic, computers and TV screens, and the stress of noise and rush, our spirits eventually become impoverished. Contact with plants, animals and rural peace, can be a marvellous antidote. We feel enriched by the beauty and magic of the natural world – because it is where we truly belong.

Alan and Rosie Beat know all about this journey. Alan had previously worked as an engineer and Rosie as a teacher, and their first tentative steps towards self-sufficiency involved planting vegetables and keeping hens in their suburban garden, much to the consternation of neighbours. Then, in the 1980s, they abandoned urban life altogether and took a leap into the unknown, buying a smallholding in an idyllic spot in north Devon.

Since then, the Beats have been on a 25 year learning curve, gradually accumulating a vast wealth of knowledge about "the good life". Throughout that time, they have shared the journey with tens of thousands of like-minded people, writing every month for *Home Farm* magazine, then its successor, *Country Smallholding*. Those articles have covered countless topics, from hedge-laying to hens and tools to tractors.

Their 16-acre smallholding is farmed on organic lines, with sheep, pigs, hens and ducks alongside positive management for wildlife. Under their management it has been designated as a County Wildlife Site and a key Devon dragonfly site, while in 2008, the Beats were highly commended finalists in the national Future of Farming Awards run by Natural England.

Alan and Rosie have made education a key facet of their smallholding life, hosting visits by numerous school and adult groups, and running popular training courses.

The couple's children, Martin and Katie, have grown up with the commitment to a sustainable lifestyle.

Alan's first book, *A Start in Smallholding*, has been popular for several years. But this book takes things a stage further, distilling two and half decades of experience into a comprehensive guide for anyone thinking of embarking on "the good life".

One final word. Alan recently wrote a feature for *Country Smallholding* entitled The Flow. It examined a remarkable phenomenon, where the unexpected happens to support what you are trying to achieve. When Alan first heard about this, he was highly sceptical. But over the years, he has witnessed it himself numerous times. He explains it this way: "When you seek to achieve something that is right for you, and right for something greater than yourself, help will flow to you in a way that defies rational explanation."

If you, too, are a seeker on the self-sufficiency journey, Alan and Rosie will make excellent guides. And you, too, are likely to taste the magic along the way. Good luck, and may The Flow be with you!

Introduction

"The wealthy and great
May roll in their state
I envy them not I declare it
I eat my own lamb
My chicken and ham
I shear my own fleece
And I wear it"

National statistics show an increasing drift of population away from areas of urban sprawl towards more rural parts of the UK. Among the many factors that are driving this trend, the search for a better quality of life is foremost. People are seeking to regain control over their lives, to grow their own wholesome food, to reduce driving on congested roads, and to raise their families in a higher quality environment. That's why more and more of us want to start smallholding!

We made that change of lifestyle with our two young children back in 1987, after four years of planning. We now keep sheep, pigs, poultry and waterfowl to produce most of our own organic food; make our own bread, beer and country wines; harvest our own firewood; and generally have become as self-reliant as makes sense to us. In the process, we have discovered new resources within ourselves. This is the essence of smallholding.

It wasn't easy to move far away from family, friends and the familiar surroundings in which we had been brought up. The culture shock of arriving in a small rural community was very real and it took time for us to adjust.

At first, nearly every aspect of our new lives required knowledge and skills that we did not possess, so we had to acquire them. New areas of interest arose from our adopted lifestyle, including friendship and cooperation with other new smallholders and established farmers. There were undoubted health benefits from the move: we became physically fitter and stronger through regular outdoor exercise and manual work, while our new direction rewarded us with a positive sense of achievement. Each day was full of interest; we loved every minute and there were few regrets.

Many years later, we have learned so much and yet we are humbly aware of how much more there is yet to learn. It seems that one lifetime will not be long enough. Years ago, we were puzzled by a remark of the elderly John Seymour, when he qualified an incident in his youth with the phrase "when I was even more ignorant than I am now". Why would a man who had spent many years living, studying and writing about rural self-sufficiency, describe himself in such a derogatory manner? Only now do we understand the truth and humility with which it was written.

Looking back, it seems astonishing that we could ever have been so ignorant ourselves. How had we gone through the education system to degree level, then lived, worked and played into our late thirties without knowing how to feed, clothe and house ourselves? All that society had really taught us was the intellectual means to earn money, without which we were helpless.

The net result of doing more for ourselves, instead of paying others to do it, is twofold. Firstly, less money is needed, so less time needs to be spent earning it, thus neatly reversing the vicious circle in which modern society traps us. Secondly, it inspires self-confidence, the belief in our ability to cope. We see this quality in many smallholders whom we have come to know; not usually overconfidence, just an unspoken inner strength.

Of course, living on a smallholding is not idyllic all the time. There are days when the weather is against us, when nothing seems to go right, or when our best efforts fail to achieve their purpose. But such hardships serve only to heighten appreciation of the better times, those other days when smallholding really is sheer pleasure. Overall our quality of life is now so different, so vastly improved from before, that it's difficult for us to imagine living in any other way.

We hope this book both inspires and empowers you to make your own smallholding dream come true.

Above: Wall trained pear tree
Opposite: an allotment sized garden

Chapter 1

First steps - the question of scale

What exactly is a smallholding? How much land will you need, or be able to manage? Many people start on the smallholding ladder as we once did: by learning to grow food crops and to keep small livestock like poultry in their existing gardens.

How big does the self-sufficient garden need to be? Our kitchen garden today measures 288 square yards, very close to the standard size for an allotment of ten poles, or 300 square yards. This area provides nearly all the vegetables and soft fruit for a family of four. Of course many modern gardens are much smaller than this, but much can be still be achieved in a smaller space, and you may be able to rent an allotment nearby, or take over an unused garden from a neighbour.

Plants need light more than anything else, so ideally you need an unshaded site that receives direct sunlight from the south. Fertile soil is a good starting point, but by no means essential, for good crops can be grown irrespective of soil or artificial surface by constructing raised beds, edged by wooden boards and filled with a mixture of rotted manure and good soil, or compost from your local community garden waste recycle scheme. In this way, any garden receiving a decent amount of sunlight can be transformed into a productive space.

Cane, bush and tree fruit can make the best use of any wall or fence with a southerly aspect, trained along wires or other supports to cover the available surface, capturing the sun's energy and transforming it into delicious food. The vertical space of a garden is then utilised for very little ground area occupied by roots. We have a single wall-trained pear tree that yields more fruit each autumn than we can cope with.

Without a suitable wall or fence, how much space does soft fruit need in the open ground? Our garden has: a twelve foot row of autumn raspberries; a strawberry bed measuring eleven by fourteen feet; six gooseberry bushes; and nine currant bushes of red, black and white varieties. In total, these occupy only a small part of our allotment-sized plot, but we harvest as much during the season as a family can enjoy fresh from the garden, plus a healthy surplus for storage and wine making.

Garden livestock

What about livestock on a garden scale? The best option is chickens. Where space is really tight, a few layers or meat birds can be kept within just a few square yards, and this is far more space than any industrially farmed bird is allocated! Be aware that chickens don't mix well with vegetable or flower gardens. Their constant scratching and pecking can inflict a great deal of damage to tender plants in a short space of time. However they may usually be allowed access to grassed areas, shrubberies, orchards and woodland without problems. For our first attempt before moving, we fenced off a corner of our back garden beneath some trees, where the hens scratched happily amongst the leaf litter. We bought and self-assembled a henhouse of modest proportions measuring 41 by 28 inches to accommodate six birds - and with a few repairs and modifications, this remained in everyday use for nearly thirty years. Many people keep their garden birds confined within a small wire run attached to the henhouse (an ark) that can easily be moved to fresh ground at intervals.

A few table chickens are easy to rear on a backyard scale. The progression from day-old chick to oven-ready bird takes around three months. The important point is to buy birds that are specifically bred for the table, as distinct from egg layers which are generally too lean. You need a small nursery area to keep the chicks warm, dry and safe from predators, including cats and rats. This could be in a utility room, garage or garden shed. As they grow, they can be allowed more room, inside or out. If moved outside, they will need housing for shelter in wet weather and for shutting away safely at night.

Rabbits are also worthy of consideration on a garden scale. We have never kept these ourselves, but have seen it done using a bank of hutches and small runs in which the breeding adults and growing offspring yielded a steady output of meat for the household from a very small space indeed. Bees can be kept almost anywhere and will forage over a three mile radius of the hive.

Opposite above: Laying hens
Opposite below: Meat chicks under a heat lamp

14

To summarise, almost any garden can contribute towards a household's annual food requirements, including vegetables, fruit, eggs, honey and some meat. We might describe such activity as microholding rather than smallholding, yet this can take you quite a long way towards self-sufficiency in food, and provide a sound basis of experience for any future move to a larger scale operation.

Moving up the scale

Moving up the scale a little, a spacious garden or small paddock offers the possibility of keeping more species of livestock such as geese and pigs, or planting an orchard for top fruit. A modest orchard of 10 to 15 carefully-chosen trees will start cropping within a few short years, and thereafter should yield enough apples, pears and plums to keep the family well supplied.

Ducks obviously need water but soon make a mess of a small garden pond set into the ground, so are better provided with a child's paddling pool or similar shallow, lightweight container that can be emptied, cleaned and moved on a regular basis. Free range your ducks if possible on grassed areas, where they will find some of their own food and be healthier for the exercise. Keep medium to large breeds out of the flower borders and vegetable garden with a low barrier or fence, for what they don't eat, they will damage by trampling. Night accommodation is essential and you may be able to adapt an existing shed or outhouse to suit.

Geese are grazers and need a decent area of grass to feed on. This needs to be short, fresh growth, for they will not thrive on coarse stuff; but at the right stocking rate, they will keep the grass mown short and save you the trouble of cutting it. In doing so, they find most of their own food and need little or no supplementary feeding from May to September, making them cheaper to keep and to fatten than other domestic birds. Goslings hatched in April can be killed in September as the traditional Michaelmas goose, or fed a supplement of cereals to keep them growing until Christmas. Geese have similar requirements for housing as ducks, scaled up to suit their increased size.

The pig was once the traditional occupant of the cottager's sty, but these intelligent animals deserve better than the dark, cramped prison to which they were often confined. At least part of their food will be green waste from the garden. The resulting manure, well rotted, is the best possible fertiliser to return to that garden; and the pig will even dig the garden for you with his remarkable snout.

You must have at least two pigs together, for no social animal should be kept in isolation. Their need for summer shelter is simply a roof over a nest of dry bedding, easily improvised with pallets, sheets of waterproof material, a few fencing stakes – whatever is to hand or can readily be obtained. Traditional breeds do well on free range and will root up the ground, clearing brambles, nettles or other rank weeds, so are best restricted to an area that can be sacrificed to their activities. They soon learn to respect electric fencing and stay confined within it.

Opposite: Geese are grazers

Outdoor pigs with ark and electric fence

However, over the cold and wet winter months, pigs will quickly transform all except free draining soil into a quagmire. The most practical options are to house them in an outbuilding with hard standing, or to avoid keeping pigs at all during this period. Eight week old weaners bought at the end of April will reach slaughter weight for pork in September and for bacon by the end of October, thus avoiding the need for winter housing while supplying enough pork and bacon for the year.

Moving further up the scale

Moving up the scale a little more, for grazing livestock such as sheep and cattle you need at least enough land to support them through the growing season. Young stock bought in the spring can be fattened for sale or slaughter in the autumn, thus avoiding the need for housing and supplementary feeding over the winter months. A few bottle-reared orphan lambs might fatten on half an acre; but if you were to repeat this on the same small patch of ground each year, parasite and disease problems would soon build up.

Grazing is best rotated between different species of livestock, or rested and "cleaned" by taking a hay crop. In practical terms, two acres or more can be sub-divided into paddocks, which enable sound principles to be followed. Hay enables some stock to be kept over winter, so females may be kept for breeding. Good pasture is reckoned to support five breeding ewes and their lambs per acre, but we suggest half that figure to start with, until you can assess the carrying capacity of your land by experience.

It's often said that a sheep's sole aim in life is to die, but we find that to be an unfair exaggeration arising from poor shepherding. Sheep may be the most difficult of farm livestock to keep and breed successfully, but nevertheless are widely kept by smallholders for their many rewards: the meat of lamb, hogget or mutton; the annual harvest of wool; milk from specialist dairy breeds; and the means by which to manage large areas of grassland.

Sheep that are trained to follow a feed bucket are so much easier to move, handle and care for than wild-as-the-hills animals that prefer jumping fences to being constrained. A two year old ewe feeding her first lambs has already proved her worth and is a popular first choice for the novice. As a minimum workload, expect to visually inspect your flock daily, trim their feet at six to eight week intervals, worm and vaccinate occasionally, and lamb and shear annually – but it's rarely that straightforward! Caring for goats is similar to sheep in these respects.

Cattle soon grow into large, heavy beasts needing stronger gates, handling facilities and winter housing than other livestock. A placid house cow might be easier to handle than boisterous beef steers, but she will of course need a milking parlour; and for a year-round milk supply, two cows are needed, calving at different times. Friends have successfully kept a house cow on three acres of pasture by alternating the paddock cut for hay each year, and housing her in an old stone barn over the winter.

An alternative to owning cattle is to rent out part of your grazing to a neighbour. This is what we do, grazing half the land with our sheep flock, and half with beef steers or in-calf heifers for the summer season. Next year, the sheep swap over to follow the cattle of the previous season. In this way, the parasite burden of the pasture is greatly reduced, for most parasites of each species are killed through ingestion by another.

A considerable benefit of housing livestock at some point of the year is the manure that accumulates. Our sheep, pigs, poultry and waterfowl combine to yield several tons of rotted manure each year. This is the powerhouse that drives the organic gardens, without which some other equivalent source of fertility would need to be imported.

Where ground and weather conditions are suitable, crops such as cereals or roots can be grown on a field scale for your own consumption or to feed livestock. Our own land is not suitable for this so we have no direct experience to offer. Other authors have suggested that for complete self-sufficiency in food including dairy products, a family of four would need five acres of reasonable land, farmed really well. This allows two acres for cows, two and half acres for cereals and half an acre for fruit and vegetables.

Chapter 2

Planning

A major change of lifestyle deserves careful planning and forethought. As you gain experience by growing food and keeping small livestock on a garden or allotment scale, it makes sense to start gathering as much information as possible that will inform the choices you need to make.

Make a point of visiting some of the country shows and rural events that take place across the UK. Many exhibitors and stallholders are knowledgeable in their chosen sphere and will usually be pleased to answer reasonable questions. We well remember visiting our first major smallholding event long ago, where there was so much to experience that, to us, was novel and exciting. For example, we watched spellbound as a sheep was milked by hand (when it simply hadn't occurred to us that sheep could be milked) and were delighted to be handed a deliciously warm, sweet sample to taste! Such a day can inspire and inform, opening the door to unforeseen opportunities.

Question anyone whom you know, or happen to meet, from a farming background. We recall speaking to a successful large-scale farmer whom we knew socially at that time. He advised us to buy more land than we thought necessary, for it is rarely possible to buy adjoining land in the future. The excess could be rented out initially to provide some income, but he predicted that we'd soon want to use it ourselves. Cynics might argue that this is merely an extension of the familiar rule that all things expand to fill the space available! But the point remains that ownership of adjacent land is the most practical way to enable future expansion of your smallholding plans.

Opposite: Ewe with lambs
Below: A country show

Learning about pigs

There is a wealth of written material to guide you along your way, led by specialist magazines such as *Country Smallholding*. Make good use of your local public library, which should carry a selection of relevant books, while any title that you fancy can be obtained on request. Internet research will help you to focus on the books that best suit your individual needs. Read as much as you can, for it all helps to build up a more accurate picture of where you are heading.

A number of specialist internet websites, networks and forums exist that cater for downsizers, greenshifters and smallholders. As always on the internet, you need to search through to find anything of direct relevance, but it's usually there somewhere, so persistence is the key.

Training and experience

Some agricultural colleges, educational charities and private individuals offer training courses in smallholding skills. These can provide a sound basis for learning, plus the opportunity to meet with other people who are following a similar path. Be aware that agricultural colleges will focus more on commercial considerations and profit margins, compared to a more holistic approach on husbandry and welfare that

Trimming a sheep's hoof

can be expected from practicing smallholders. Course fees, length and content vary considerably, as does the expertise of the tutor, so do your homework thoroughly to find the course that is right for you, by internet research or through advertisements in smallholding magazines.

Beyond that, there's nothing like a taste of the real thing through hands-on work experience. World Wide Opportunities on Organic Farms (WWOOF) is a unique organisation linking volunteers to an international network of small farms offering board, lodging and training in exchange for your labour. Our two children have gained experience on farms and smallholdings across the UK and Europe through this marvelous organisation, and both thoroughly recommend it. There's a solely internet-based network called HelpX that could also be considered. No two smallholdings are the same, so there is something to learn from each one visited.

Pitfalls

There are common pitfalls to avoid during the planning process and it helps to be aware of these:

• Don't just visit smallholdings in the summer months and gain the false impression of a sunlit rural idyll; go in the winter as well to experience the mud, rain and cold, for an all-year-round picture.

• Don't decide which livestock you will keep without experiencing and handling them, or you may find that you are unsuited.

• Don't set your heart on starting with pedigree, show-winning livestock, despite this flying in the face of advice given to beginners. These are the most expensive and difficult of livestock to keep and breed well, so reserve this option for later.

• Do settle for gaining experience through keeping commercial or cross-bred livestock, which are generally hardier and easier to manage; and should things go badly wrong, the costs involved are much lower.

• Don't get carried away by your own enthusiasm or that of others. Keep both feet on the ground; be realistic about how you will cope with the physical demands of smallholding; and be realistic, too, about how much income the smallholding will contribute to your budget. On the other hand . . . .

• Don't be a pessimist. It's important to take a positive view of things; is the glass half full, or half empty? To the pessimist, obstacles are insurmountable so there's no point in trying. To the optimist, obstacles are a challenge to be overcome, in the process of which something useful may be learned to make life easier next time. Most things in smallholding are achievable if you have a positive outlook.

• Don't rush into anything. Take your time, do the research, build up gradually by taking over more of the garden or an allotment, before moving on to a larger scale operation. We spent four years in this phase and it was time well spent.

Opposite: The kitchen garden in winter

Above: An idyllic country cottage
Opposite: A smallholding with traditional outbuildings

Chapter 3

Finding a smallholding

Where should you look for a smallholding? There are many personal choices involved in answering that question. Some look abroad to Europe, where property prices are generally lower and the climate may be more favourable; but against this must be weighed the barriers of distance, language and culture. Even here in the UK there are considerable regional variations in climate, geography and property prices, while some cultural differences also remain.

Many variables need to be considered before you can start narrowing your search to a general area. Look carefully at weather patterns across the UK, particularly temperature and rainfall, as these dictate the length of the growing season and which crops will thrive without protection. The websites of estate agents can be surveyed for availability and price of suitable properties, although a virtual tour is no substitute for a site visit. Weekend breaks or longer holidays can be used to explore an unfamiliar area, although it must be borne in mind that first impressions run the risk of being overly optimistic. A close look at the farmed landscape will at least give an idea of which crops and livestock are best suited to the area.

Distance from family and friends is another factor to consider. Paid employment is a necessity for many would-be smallholders - can you work from home in this technological age, or are suitable jobs available where you hope to find a smallholding?

A smallholding on marginal land

Narrowing the search

Choosing an area of the country to live in is one thing; but finding your dream smallholding within this can be far more difficult. You'll need to visit and explore as much as possible to build up a feel for the area, gradually narrowing your search to those parts that best match your aspirations. It's worth keeping in mind that geographical features like altitude, aspect and soil of any holding are fixed; whereas the house, outbuildings, field boundaries, access and drainage can all potentially be changed, albeit at a cost. It helps to draw up a list of your requirements and to judge each potential property against these, rather than just vaguely liking the house or the view. A very effective approach is to temporarily rent a property within your chosen area, enabling you to search from a local base.

As your search narrows to individual properties, the crops or livestock that are currently supported on the holding can be compared against those you hope to introduce. Soil samples can be collected and analysed to eliminate guesswork. New buildings may require substantial investment, so any existing outbuildings are important features to be assessed with a view to their use or adaptation to suit your needs. Evaluate whether gates, fencing and handling facilities are adequate to contain

Outbuildings can be renovated, at a cost

your intended livestock, and if they are not, estimate the costs of upgrading accordingly.

Buying or renting

Perhaps the most important advice of all is to rid yourself, if at all possible, of the mortgage on your home. If you can, it will make a real difference to the financial viability of your new life in smallholding. When we made our own move more than twenty years ago, this was central to our plans and a crucial factor behind the decisions that we made. We found that suitable smallholdings were unaffordable in our original target area, but came within reach as our search shifted further west, away from major cities and transport links. Of course, property prices have changed dramatically since then, but all things are relative, so you still may be able to achieve this through careful choice. For example, you might decide to sell your existing home first and move into rented accommodation within your chosen area, wait until the right property enters the market, and then negotiate from the strong position of being a cash buyer.

Smallholdings may be available for rent but these are few and far between and you'll need to work hard to find one. Try a wanted advertisement in local newspapers and on specialist websites, and make extensive local enquiries within the area. There'll be far less choice so you'll need to be flexible. Beware of buying or renting land that is separated by any distance from the dwelling. Ten yards away across a country lane may be fine, but two miles away is a significant barrier. Many people try this route into smallholding, but most find that the chore of travelling to and fro, and the difficulties that arise from not being present at critical times, eventually force the move to a self-contained smallholding.

Beware also the trap of buying a bare piece of land in the UK and expecting to build your dream house upon it. This is possible in theory, but very difficult to achieve in practice. The planning authorities will need convincing that you can earn the minimum agricultural wage from your holding, and that you need to live on site to deliver this. Most smallholding activities will fail on both counts, so you really do need to research carefully and to seek professional advice before committing to this approach.

Community farming

Another route is to consider joining a community, or setting up a new one. These can take many forms, and one that we know involved the collective purchase of a small farm that could house several individuals and families. A number of people expressed interest in the idea, meetings were held and various possibilities explored. If a way could be found to finance the project collectively, there was the potential to enjoy the use of more land, outbuildings and facilities than could be financed individually. The project was then established by 11 residents, who were equal members of the limited company that bought the farm. The farm is now a co-operatively run eco-village, with each resident having a private living space as well as shared use of communal areas. The community aims to live in a low impact, sustainable manner and to approach self-sufficiency in most things; this is smallholding on a community scale.

Above: Check for stockproof fences
Opposite: A community farm

Above: A productive vegetable garden
Opposite: Raised beds edged with wood in a small garden

Chapter 4

Growing

A very efficient way of growing your own organic vegetables and fruit is to use deep or raised beds dressed with heavy surface mulches. By working with nature rather than against it, digging and hand weeding are minimised while yields are maximised. The beds need not be raised or edged, but crucially they are not walked on and are worked instead from the paths on either side to avoid soil compaction.

Most people find four feet a comfortable width for the beds, with paths two feet wide for wheelbarrow access and for kneeling. Ours are aligned north to south to ensure even distribution of sunlight. To establish deep beds without edgings, double digging is beneficial to loosen and aerate the soil and enable manure to be worked in, but you will only need to do this once.

For raised beds, no digging is really necessary, though instead you have the work of constructing the edgings in wood, brick, stone, slate or other materials. We have reservations about using chemically treated timber in the garden, so our preference is to recycle scrap natural roofing slates, pressed vertically into the ground and overlapped slightly to form a continuous barrier. These take up no growing space, are cheap or free to obtain, sustainable, aesthetically pleasing, effective and durable. Their drawback is that the height of the bed is limited by the size of slate available. Paths are easily kept weed-free by laying down old carpet or black plastic, and covering with wood chips if desired to soften their appearance. A garden laid out in this fashion is pleasant to look at, accessible all year round and potentially highly productive.

Emptying the muck heap

Muck and mulching

To maintain and improve productivity, all soils need feeding, and the best source of nutrients is decomposing organic matter of one sort or another. Compost is easily made by recycling green waste from the kitchen and gardens to produce a valuable fetiliser and soil conditioner. Well-rotted animal manure is a beneficial side-product on smallholdings with livestock, and can be the powerhouse that drives organic food production in the garden.

We don't make compost as such, but instead recycle green waste through our pigs and thus to the muck heap. Every autumn the muck heap is emptied onto the garden and most beds receive a surface layer three to four inches thick, straight on top of any weeds or crop residues, for very little will grow through it. This continues to decompose over the winter, releasing a trickle of nutrients to be washed by rain into the soil beneath. Meanwhile the worms attack it from below, gradually incorporating it into the top layer of soil, aerating this with their burrows and fertilising it with their casts.

In the spring, large seeds like peas or beans are sown direct into this mulch layer, or young plants raised in the greenhouse are transplanted from their pots straight into

Onions grown in a surface mulch

matching holes cut with a bulb planter. The surface texture of this mulch is such that few windblown seeds will germinate in it, so the crops grow on largely without competition from weeds. The few weeds that do manage to gain a foothold are easily pulled by hand.

Most crops are harvested by cutting or pulling to leave the surface layer largely intact, so it will continue to suppress weed growth through the winter if required. By the following spring, eighteen months after spreading, the layer will have reduced to a fine, dark, friable compost that is just right to receive direct sowing of the small seeds of salad and root vegetables. Generally speaking, salad and root crops require less feeding than other vegetables, so this approach works out well.

You can see from this that by planning ahead, most of the vegetable beds remain covered by mulch and weed-free for most of the time, so there's far less work involved than with bare soil. Against this must be balanced the labour of moving and spreading the muck each year, although most conventional growers would shift this anyway, and then dig it in as well!

Waste cardboard, junk mail and other paper products can be recycled through compost heaps, but we recycle it directly onto the ground around fruit bushes and trees, overlapping the sheets to exclude any light and then covering with a thick layer

of old hay, straw or rushes to hold them in place. This kills weeds and deters germination as the mulch gradually breaks down. After twelve months, as weeds finally start to show, we put down a fresh layer to start the process again. By this means the ground between fruit bushes and trees is kept largely weed free and fed at the same time. Avoid using glossy magazines and coloured printed matter, as the inks used can contain hazardous chemicals. Also avoid window envelopes, or first remove the plastic window inserts.

Wood ash from enclosed stoves or bonfires is a valuable source of potash so has particular benefit for soft and top fruit. Sprinkle it around bushes and trees for rain to wash in. A better use of wood ash is to spread it beneath the perches of a chicken house, where it reacts chemically with the fresh droppings of poultry to neutralise smell and produce a brownish substance that has more value as a fertiliser than either constituent on its own, and can be used straight away.

Pest and disease control

Crop rotation prevents the build-up of pests and diseases that would occur if the same crop was repeatedly grown in the same spot. Plants of the same family are grown together and move on to a fresh location each year. Our garden has six main beds so we practice a six year rotation, simply moving everything along to the next bed each year. Heavy feeders like brassicas and potatoes receive the manure surface dressing, then salad and root crops follow a year later when the mulch has composted down.

Pest control in the organic garden is largely a matter of finding the right balance between pests and their natural predators. It is about control, not elimination, for without some pests you cannot retain their predators. Hedgehog, slow-worm, toad, frog, thrush and many small birds are your allies, encouraged by perching and hiding places around the garden such as bushes, log piles or old clay pots. Lacewing and ladybird "hotels" encourage the presence of beneficial insects in the spring and early summer when they are most needed. A mixed flower bed in the vegetable garden attracts a range of insects, including pollinators and pest predators.

Opposite: Surface mulch around fruit trees
Below: Ladybirds are beneficial predators

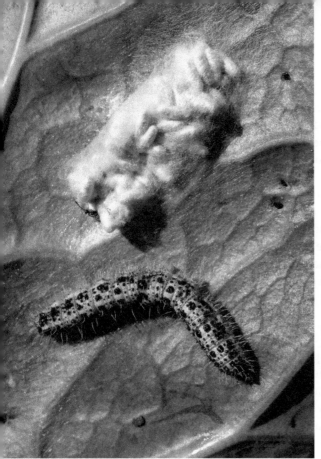
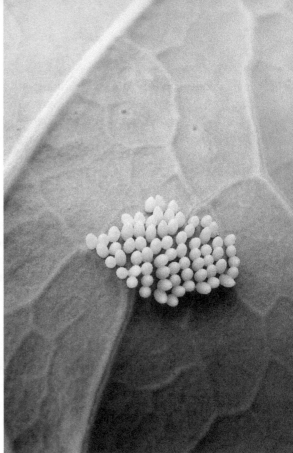

Above left: Caterpillar (top) consumed by encarsia
Above right: Butterfly eggs on cabbage
Opposite: Lifting new potatoes

Watch for the bright yellow eggs of butterflies on the underside of brassica leaves to squash them before they hatch, and pick off caterpillars by hand as soon as you see them, but be careful to leave any that are lethargic and clearly unwell by comparison to the others; for these may be affected by the parasitic wasp *encarsia* that are your friends.

We keep a flock of call ducks specifically for slug and snail control. These very small ducks patrol the gardens for part of each day, actively seeking these pests among the plants. The little damage that they do is far outweighed by the benefit. When asked about slug control, we answer, "If you have too many slugs, you don't have enough ducks".

Minimum-dig vegetables

In our system, potato seed is dibbered or trowelled straight through the surface mulch into the soil beneath to a depth of around six inches - and that's it; no earthing up, no weeding. There are usually a few green tubers at the surface to discard at harvest, but otherwise a good crop can be expected. For the very first earlies, we set a

Above: Leeks following early potatoes
Opposite: Runner beans

few seeds under cloches made from old window frames, to protect against frost and to warm their growth ahead, removing the cloches once foliage fills the space within. At harvest, the remaining mulch layer is forked into the top spit of soil as the tubers are dug out. This is the only real digging that our beds receive, once in every six year rotation.

Onion sets and shallots are pressed straight into the surface mulch leaving just the tip showing. Cover these with chicken wire to deter birds from pulling them out, and then remove it once the bulbs have rooted firmly.

Leeks are sown in large pots or a seed bed in open ground, then separated as bare root transplants to follow early potatoes. Make a deep hole with a dibber, pinch off about half the root length, drop the young plant into the hole and fill with water, but not soil. There's some hand weeding to do later because the bed has been dug for the preceding potato crop, bringing weed seeds to the surface.

Brassicas are sown in pots and transplanted when they have grown six to eight leaves. Cut a neat hole with a bulb planter to fit the root ball and pop it straight in, down to the level of the seed leaves before firming. Cover with a two litre plastic bottle cloche (base cut off and screw cap removed) and water. The plants are scarcely checked

Above: Brassica transplants
Opposite: Vegetable harvest

by this method and grow on strongly. Remove the bottle cloches as the space within fills with leaves. Stake tall brassicas like sprouts and broccoli to prevent wind rock during winter gales.

Legumes have large seeds that can be dibbered straight into the mulch, but tender crops like climbing French and runner beans are best sown in pots for transplanting after the risk of frost has passed. These quickly outgrow slug damage by racing up the canes. A week later, make a second sowing directly into the mulch layer, to give a later harvest of beans into the early autumn. Broad beans from a November sowing can be overwintered under window frame cloches to give an early crop the following May.

Salad and root crops are small seeded and generally give most work to the minimum-dig gardener. The best way of sowing is directly into the fine compost remaining from the surface mulch of the previous year, drawing shallow drills and covering the seed as usual. Germination is usually good, but hoeing or hand weeding will be necessary because the weed-suppressing qualities of the mulch have passed. Where crop rotation demands a different approach, furrows can be formed in the manure mulch by dropping a heavy metal bar down flat on the surface. Partially fill the depression with fine compost, sow the seed, then cover with compost. It's a bit fiddly, and slug damage tends to be higher.

Above left: Cutting asparagus
Above right: Pruning a new blackcurrant bush

Rhubarb and asparagus are permanent crops that are well worth growing, despite taking two years to establish, for their harvest comes in the spring when other fresh produce is scarce. These fall outside the general crop rotation, and are generously mulched with muck each autumn to suppress weeds and feed next year's crop.

Herbs

A natural component of the kitchen garden is an area for growing herbs, especially those perennials that complement the food you produce; mint with new potatoes and peas, rosemary sprinkled on roast potatoes or a joint of lamb, and chives chopped into a potato salad.

Mint is rather invasive so grow it with roots contained if possible. Cold summer drinks are lifted with the addition of a few mint leaves. Spearmint and apple mint have the best flavour for culinary use. Rosemary and sage both like a hot, sunny situation and are best cut back in the spring to keep them from becoming too woody. Use fresh sage with onion for stuffing chicken and turkey.

White and red currants

Chives are easy to grow and do well as an edging in the garden. They are very decorative and the flowers can be added to salads along with nasturtiums and marigold petals. Thyme is another herb that is useful in stews and stuffing, and for Italian pasta dishes you'll need a clump of marjoram (oregano) too. Coriander and parsley are best treated as annuals, with fresh seed sown every year. Make room for a few plants of basil, a must when preparing tomato dishes, but you may struggle to keep slugs away from these.

Soft fruit

Soft fruit bushes and canes earn their place in the self-sufficient garden by yielding a delicious annual harvest in return for little work. These are permanent crops that take two or more years to establish. Buy certified virus-free stock and space six feet apart, working plenty of muck into a generous planting hole.

Blackcurrants are pruned back hard within two inches of ground level straight after planting; this seems like harsh treatment, but in fact promotes the vigorous growth of new canes that will fruit in their second summer. After that, progressive renewal is encouraged by annually removing one third of the fruited canes, plus any weak or badly placed new shoots.

Above: Strawberries
Opposite: Dessert gooseberries

Red or white currant and gooseberry bushes are pruned to establish a permanent framework of open branches supported on a single stem. Traditionally this work is carried out in the early winter after leaf fall. Summer pruning can also be used to contain the growth of vigorous bushes and to enable light and air to penetrate and assist the ripening process. Where space is tight, most soft fruit can be trained along support wires against a south-facing wall or fence and pruned accordingly to establish the desired shape.

Our fruit canes, bushes and trees are treated to a surface mulch each year, either of waste paper and cardboard covered with hay as already described, or a feed of rotted manure every third year or so. They are also fed regularly with wood ash. Strawberries are tricky because of their low-growing habit, but can be planted initially into a deep surface mulch, which is then topped up each year between the plants. There is some hand weeding most years but we can live with that for the delicious harvest. Every few years, vigorous runners are selected to start a new bed in a fresh location.

We no longer grow summer raspberries, finding that the yield declines after a few years until the work involved is no longer justified. Instead we favour the autumn varieties like Autumn Bliss, which yield well over many years and are easy to grow; just cut down the old canes in February to ground level, and loop strings around the outside of the new canes as they grow for mutual support.

Birds will decimate soft fruit unless kept away by a fruit cage or temporary netting, draped over and held in place by pegs or weighted down by lengths of wood. Wasps can also be pests and in bad years may attack in their thousands to hollow out the fruit before it ripens. If we see this problem building up, our solution is to track the offending insects to locate their nest nearby, and then to return after dark to knock it out by firmly blocking the single entrance hole to suffocate the inhabitants.

Top fruit

Fruit trees take time to establish, but you'll be enjoying the first fruits within a year or two of planting, so it's well worth planning your orchard straight away. Alongside modern favourites, consider local varieties that are best suited to the area, and plan to spread the harvest over a long season, including some apples that store well into the New Year. Select trees that are well shaped, and plant semi-dwarf rootstocks spaced twelve feet apart. Work plenty of manure into a generous planting hole, and stake for support on exposed sites. Protect the young trees from livestock, including geese, which will ring bark and kill them, as we have learned to our cost.

Most fruit trees do well against a south facing wall, enjoying the shelter and reflected warmth of the situation. Pear blossom opens earlier than most fruit, so really benefits from frost protection. We planted a Williams bon Chretien in such a situation and trained it informally up the wall, nipping out the leader point to promote side shoots each year. After three years we tasted our first pears; and within a decade the tree was yielding over three hundred fruits each autumn (photo on page 12).

Pruning is carried out in late winter just before bud burst to minimise the risk of disease entry via cut surfaces. Aim for an even structure of four or five main branches on freestanding trees, keeping the centre open and removing any crossing growth.

Apple harvest

Most trees carry fruit on short spurs, and these are promoted by pruning side shoots back to four or five buds, then pruning again the following year to the topmost flower bud. Once the spur system is fully developed, little further pruning is required. Summer pruning can also be used to restrict growth of the tree and encourage ripening of the fruit.

Wasps have been a serious pest of our top fruit and are particularly attracted to the early apple varieties as they sweeten towards ripeness from late July onwards. As already mentioned, dealing with nests nearby is the best from of defence. At times we have successfully protected fruit by wrapping individual clusters within horticultural fleece, although this is rather fiddly and time consuming. In the past grey squirrels have occasionally damaged our harvest by pulling unripe fruit from the tree, taking a single bite and throwing it to the ground. These can be dealt with by trapping, although our preferred solution is simply to keep a cat!

Call ducks in a small garden

Chapter 5

Poultry and waterfowl

Poultry and waterfowl can provide a regular supply of eggs and meat for the kitchen or for sale. They can hatch eggs and rear chicks; help to control garden and field pests; and in the case of geese, act as guards to alert you when people or predators approach. Nor are they difficult to look after. Buy healthy birds from a reliable source; feed and house them properly; keep them free range or move enclosed runs regularly; and you shouldn't have any problems. A basic requirement for all birds is to shut them safely away from predators before nightfall and let them out again in the morning.

Hens

Laying hens are often bought as point-of-lay pullets, at around 20 weeks old when egg laying is about to start. Their first eggs are small but soon increase in size. Birds that start laying in the autumn should lay right through their first winter until the first moult in late summer, when egg production ceases for several weeks while new feathers grow. Egg laying then starts again, but at a lower level during the second winter as production is linked to hours of daylight. Overall production declines progressively as the bird ages. These natural traits have been taken to extremes in some modern hybrids that are designed for maximum indoor production in their first year of lay and then discarded. From the smallholders point of view, these specialised birds are smaller, less hardy, and often short-lived by comparison to more traditional breeds and crosses, or modern hybrids that have been developed for free range rather than caged conditions.

Housing must be rat, fox and weather proof, with ventilation towards roof level but otherwise free of draughts. Small outbuildings can be adapted, ready-made poultry housing is widely available for purchase, or the resourceful smallholder can make a small house quite easily from recycled materials. There should be perches for roosting and nest boxes for egg laying. Perches should be higher off the floor than the nest box entrance, otherwise birds will roost there instead and foul it. Position multiple perches at the same level to discourage competition for the highest one. Nest boxes are most used when positioned low down in the darkest part of the house and lined with inviting nest material such as clean straw or wood shavings.

Droppings accumulate beneath perches overnight and need to be cleaned out regularly. As mentioned in the chapter on growing, wood ash spread on the floor of the poultry house reacts chemically with these droppings to neutralise smell and form a valuable fertiliser. Food can be provided in self-help hoppers but these inevitably attract wild birds and rodents. We prefer to feed a measured ration outside the house

every morning that is cleared up within a few minutes; if any remains uneaten, reduce the ration accordingly. In the evening we feed a second ration within the house, attracting the birds inside for the night. Again this should be entirely consumed before the birds perch up. You soon find the right balance by trial and error, and this partly depends on how much natural food the hens find whilst foraging during the day.

Organic layers pellets are now widely available and are a complete compound feed. However hens are omnivores and thrive on variety, so we also feed mixed poultry corn. Birds without access to grass should be provided with fresh greens from the garden on a daily basis. Drinking water should be available at all times in simple recycled plastic containers.

Hens also need dust baths to express their natural behaviour for controlling external parasites. Free range birds will find their own places, but a box of fine, dry soil or sand should be provided within an enclosed run.

Hens originate from jungle fowl so have an instinctive dislike for wide-open spaces, preferring the shelter and shade of trees and shrubs. Try to accommodate this when siting the hen house and range areas. Never enclose poultry in the hot sun without shade.

Free-range poultry are at risk from foxes, which may attack even during the day when feeding cubs or unable to catch more difficult prey. Years may pass without incident, but sooner or later it will happen. The occasional loss can be accepted philosophically, but at times we have taken action against a persistent daytime attacker, either by temporarily enclosing our poultry, or by dealing with the individual animal. Dogs, noise or human activity close by are good deterrents but not infallible. Electrified fencing is perhaps the most effective protection and is widely used on commercial free-range units.

Opposite: Inside a house for six laying hens
Below: Hens laying in a nest box

Young table chickens fed greens in a run

Table chickens

A small batch of table chickens is easy to rear on a backyard or smallholding scale. The progression from day-old chick to oven-ready bird takes around three months. Buy your day-olds from an organic source for preference, or from a local small-scale breeder. There are many breeds and hybrids to choose from but the important point is buy birds that are specifically bred for the table, as distinct from egg layers which are generally too lightweight.

First set up a small nursery area to keep the chicks warm, dry and safe from predators, including cats and rats. Provide absorbent bedding in the form of sawdust or shredded paper, and warmth from an infra-red heat lamp. Ensure there is space for the chicks to spread out beyond the warmed area, and their behaviour will indicate the temperature required. Too cold and they huddle tightly together beneath the lamp; too hot and they circle outside the lamp; just right and they spread across the warmed zone, stretched out blissfully like sunbathers on a beach. Adjust the temperature by raising or lowering the lamp accordingly.

Provide food in the form of organic chick crumbs. Sprinkle crumbs into a shallow container on the floor of the brooder, to keep it separate from the bedding on the floor.

Provide water in a drinker that prevents chicks from climbing in and drowning – or they will! Make one by cutting one or two openings in the side of an empty plastic milk bottle, just large enough for the head, but not the whole body, to pass through. Three-quarters fill the bottle with water and lay on its side in the brooder. Remember to scale the hole size upwards as the chicks grow.

Aim to progressively reduce the warmth until the birds are "off heat" altogether by around three weeks old. By this time the chicks will have outgrown their nursery accommodation and can be allowed more room, either indoors or outside. If outside, provide housing with clean bedding for shelter during inclement weather and in which the birds can be safely shut away at night; improvise with anything from old pallets to a small garden shed, but it must be dry and fox proof. Perches are unnecessary. Progressively change the feed to an organic growers ration around five weeks of age. For birds reared indoors, provide a regular supply of greens from the vegetable patch.

The chickens should be ready for slaughter at round 12 weeks old, though we often grow them on for another two or three weeks for a heavier carcase. Withhold all food (but not drinking water) for 24 hours prior to slaughter. Chickens are relatively easy to kill, pluck and dress at home. Find someone locally who can show you how to do it properly the first time. Deal with some of the birds yourself so that you'll know next time. Alternatively, book your chickens in for processing at a local facility (if there is one) carefully avoiding the December rush.

Ducks

Ducks obviously need access to water, and a child's paddling pool or similar container can easily be emptied for cleaning or moving to fresh ground, whereas a permanent pond will quickly be transformed into a muddy mess. Free range your ducks if possible on grassed areas, where they will find some of their own food and be healthier for the exercise. Keep larger breeds out of the flower borders and vegetable garden with a low barrier or fence; however miniature breeds, such as call ducks, do far more good than harm by actively hunting slugs and snails.

An existing shed or outhouse may be easily adapted to house ducks safely overnight. Allow two square feet of floor space per bird and provide clean, dry bedding of straw, wood shavings or similar. The perches and nest boxes of a chicken house are not required because ducks roost and lay their eggs at floor level. Most eggs are laid at night or soon after daybreak, so are collected once the birds have been let out for the morning. Ducks do not cope well with sudden changes in level, so avoid steps or steep ramps at the house entrance to reduce the risk of leg injuries.

Like chickens, ducks are easily trained to enter the safety of their house before dark with an evening feed placed inside. In the morning they will follow eagerly to wherever breakfast is provided. Contain new birds within a run attached to the house for several days until they are used to the routine (and to you) before allowing free range. Drinking water should always be available.

The Khaki Campbell is the supreme laying breed, a good strain of which will lay over 300 eggs a year; that's one egg almost every day, apart from the late summer moult, when laying ceases for several weeks. This ability to lay through the dark

Child's paddling pool as a duck bath

winter months can be used to ensure a regular supply of eggs for the kitchen when most hens are off-lay.

For table birds, choose a heavier breed like the Aylesbury and grow from day-old in similar fashion to meat chicks. The main difference is that chick crumbs must be soaked in water for a minute or two before feeding as a sloppy porridge, as small ducklings have difficulty in swallowing dry food; this can be phased out after the first seven days. Surprisingly, ducklings cannot waterproof their down or feathers until their oil glands become active at five or six weeks old, and wet ducklings soon chill and die; so deny access to bathing water until that stage is reached. Meat ducks are slaughtered at precisely eight weeks old, or otherwise older than 12 weeks, to avoid the emerging stubs of new feathers spoiling the appearance of the carcase.

Geese

Geese are grazers and need a decent area of short, fresh grass to feed on. At the right stocking rate, they will keep the grass mown short and save you the trouble of

Geese need access to water

cutting it. They find most of their own food and need little or no supplementary feeding from May to September, making them cheaper to keep and to fatten than other domestic birds.

Geese have similar requirements for housing as ducks, scaled up to suit their increased size. They have a reputation for aggression but this is mostly show. Don't be fooled into believing that a gander will protect his wives against foxes; he is just as likely as they are to be killed and carried off.

Geese lay relatively few eggs during the spring season only, so are most useful in terms of producing meat for the table. Goslings hatched in April will need a good start on chick crumbs, but this can be phased out after a few weeks. They continue to fatten on grass alone until leaf production slows in the autumn, when they can be killed as the traditional Michaelmas goose, or fed a supplement of cereals to keep them growing until Christmas. Use the services of a professional processor if you can, as plucking by hand is notoriously slow and difficult.

Other birds including quail, guinea fowl and turkeys may also have a place on the smallholding but lie beyond the scope of this book.

Chapter 6

Pigs

When William Cobbett wrote *"Cottage Economy"*, the founding manual of self-sufficiency in 1821, the pig and its products were accorded the utmost respect. He advised the cottager to avoid the difficulties of breeding his own, but to buy growing stock instead.   He pointed out that:

"all pigs will graze, and therefore, on the skirts of forests or commons . . . find a good part of his own food from May to November . . . . especially if the cottager brew his own beer, which will give him grains to assist the wash"

Cobbett scarcely mentioned pork, but instead praised fat bacon:

"Some other meat you may have, but bacon is the great thing. It is always ready; as good cold as hot; goes to the field or the coppice conveniently; has twice as much strength in it as any other thing of the same weight; and in short, has in it every quality to make a labourer's family able to work"

He finished with a practical observation that would not even occur to most modern farmers, brought up to expect fertility out of a chemical bag:

"The fatting of a large hog yields three or four loads of dung, really worth more than ten or fifteen of common yard dung (i.e. cow dung). In short, without hogs, farming could not go on; and it never has gone on, in any country in the world. Hogs are the great stay of the whole concern . . . without them the cultivation of the land would be poor, a miserably barren concern"

Today we prefer leaner meat from smaller, younger, faster-growing animals, but the outstanding usefulness of the fattening pig to the smallholder is unchanged. The pig converts his food into top quality meat more efficiently than any other large livestock, and part of this food will be garden waste (but note that regulations now prevent the feeding of any scraps from the kitchen). The resulting manure is the best possible fertiliser to return to that garden; and the pig will even dig the ground for you, or plough up and reclaim rough land with his remarkable snout.

Certainly it's difficult for us to imagine our own smallholding running efficiently without a few pigs around to mop up the waste. For a decade we kept breeding sows, though more recently we have bought in weaners each spring for killing in the late autumn. Overall we have found that Cobbett's advice broadly holds true today: for the purpose of self sufficiency in pork or bacon, it is probably better to buy in weaners than to breed your own.

Opposite: Crossbred weaners sunbathing

Crossbred weaners exit the trailer

Buying weaners

Piglets are traditionally weaned at around eight weeks old and should weigh thirty to thirty five pounds apiece, but buying your livestock at this stage is far from the beginning of the story. You need to go right back to the breeding adults and their husbandry.

Consider the worst case scenario. The sow may be dosed for parasites with systemic chemicals, which enter every living cell in her body, so must also affect her offspring. The newborn piglets may have their teeth clipped back to gum level (to prevent biting of the sow's teats), their tails docked (to prevent tail-biting) and an injection of iron administered (to prevent anemia), all within the first few days of life, while boars may be castrated (to prevent "boar taint" in the meat).

As the piglets grow and begin to take solid food, they may be fed a diet containing copper and antibiotic growth promoters that are widely and routinely used. They may be weaned much earlier than the natural time of eight weeks, right down to three weeks from birth, because this also is more profitable. They may be housed in confined pens which deny natural behaviour patterns.

I could go on, but hopefully the point has been made: however carefully you manage and feed your growing pigs, you cannot undo what has already been done. So

Tamworth weaners

if you aim to produce additive-free, humanely reared pork, you need to start with weaners that have been bred and grown to those standards.

The horrors listed above are "solutions" to problems that result from intensive management, and the small-scale pig keeper is unlikely to have any need for them. We certainly did not find the need for tooth-clipping, tail-docking or iron injection for our own piglets, nor did we castrate, having proved to our own satisfaction that so-called "boar taint" does not exist in the pork or bacon from pigs that we have reared and killed before nine months old.

The obvious solution is to buy your weaners from a good source, and to establish this you need to visit before buying. Never mind if you are a novice, common sense will be your guide. Ask to see the breeding sows and boar as well as any piglets; look at the conditions in which they are housed, and the space they have. Look at the whole litter of piglets to see what size difference there is between them; an even litter is ideal, but if there is much variation, avoid obvious runts and pick the largest. Be aware that boars will grow slightly faster than gilts (females).

Like all animals, pigs that are healthy and thriving will be bright-eyed, alert, active and inquisitive - though if they have full bellies, they may be contentedly sleeping! Look for a good level of rapport between the owner and his livestock. Ask questions about their feeding and husbandry; any good breeder will be enthusiastic and happy to answer reasonable queries. If he isn't, look elsewhere.

It is well worth establishing by your questions how the weaning process will be, or has already been, managed. Accepted best practice is to quietly remove the sow out of earshot, leaving the piglets behind in familiar surroundings and eating food to which they are accustomed. Over the next few days they are than fed little and often, at least three times daily, to minimise digestive upsets following withdrawal of the sow's milk. Once they have settled to this feeding routine, they are ready for sale. This is an appropriate point at which to worm the piglets should this be thought necessary, for instance if they have been on ground shared by other pigs in the recent past.

None of this is much to ask and will go a long way towards avoiding problems after you have taken delivery. It has to be said that many weaners are less fortunate, being simply taken straight from the suckling litter for delivery – rather less fuss for the breeder, but a much higher risk for the purchaser, so beware.

A sudden change in feed may cause digestive upset, leading to scouring, dehydration and weight loss, weakening the animal and leaving it vulnerable to other ailments. To avoid this, stick with the same feed that they were reared on, or else change gradually over several days from the old to the new ration, perhaps buying a small quantity of feed from the breeder for this purpose.

Opposite: Berkshire weaners with ark and electric fence
Above: Tamworth and Berkshire pigs

Tamworth and Gloucester Old Spot weaners in an outside run

Locating a good source of weaners may not be easy, then; and when you find a source, there may be none available at the time you want them, so start enquiring in good time. Traditional breeds offer real advantages of hardiness and flavour over modern hybrids that have been developed for rapid growth at the expense of most other qualities, including taste! Crossing two or more traditional breeds imparts hybrid vigour to the offspring, which grow faster, and are generally hardier and healthier, than pedigree stock (see the sheep chapter for more detail). You must buy two or more, for pigs are social animals and should not be kept alone.

Housing and management

If weaners are bought in the late spring when the ground is drying out and temperatures are rising, their housing and management are relatively straightforward. Free range is fine if you have an area of rough ground that the pigs can plough up, but may not be possible when you need to conserve valuable grazing for other livestock. Free range pigs need to be contained within fencing such as woven stock wire, which they are liable to uproot unless there is an extra strand of taut barbed wire at ground

The same run after a wet summer

level. They will respect electric fencing, but like all livestock need to learn first, and untrained weaners may charge straight through it.

Summer outdoor pigs are happy with a simple shelter that is easily home-made from straw bales, wooden pallets, corrugated iron or whatever is available. It needs only to be weatherproof with ample dry bedding. However, during the winter months, free range will inevitably become a mud bath unless you are blessed with exceptionally free draining soil. On heavier land, winter pigs need permanent housing in accommodation large enough to provide sleeping, dunging and exercise areas. Our solution was to concrete an outside exercise yard leading off an existing stone building, and to contain the pigs here with permanent fencing.

Feeding

Wholesome pig food, free of additives and genetically modified ingredients, is readily available today. We buy organic nuts from a local agricultural merchant, not because we need to comply with any certification scheme, but so that we know what the pigs, and ultimately ourselves, are eating. It's expensive, but we are glad to have the choice.

There's a further choice – is it better to feed ad-lib or restricted rations? Our experience has been that pigs fed ad-lib, that is as much as they will eat, grow more quickly (as you'd expect) but are too fat at slaughter. To achieve the right carcase condition, feed a measured diet right the way through from weaning, starting at one-and-a-half pounds per head per day of dry meal or nuts, rising steadily to about four-and-a-half pounds per head per day at twenty two weeks old. From that point on, feed a level ration to reach the live weight target of one hundred and fifty pounds for pork, or two hundred pounds plus for bacon. Handle the pigs regularly and learn to gauge their condition by feeling the depth of cover over the backbone; if too thin, you are underfeeding, and vice-versa.

These figures assume a daily ration of compound feed averaging 17% protein, with plenty of additional greenery on free range, or fruit and vegetable garden waste thrown into pens. Penned pigs should be given soil regularly on the roots of weeds or as a fresh-cut turf now and then, as a natural source of minerals; this is especially important for new born piglets., which can otherwise suffer from iron deficiency (anaemia). Potatoes should be cooked before feeding or young pigs will fail to digest them, just as humans do.

Weaners that are eight weeks old in May should reach pork weight in September, or bacon weight by October/November.

Below: Saddleback pigs used to clear bramble and bracken

Slaughter

Pigs need to be booked in for slaughter several weeks in advance – but first find your abattoir. The ideal is a local, small-scale, fully licensed operation. After the time and care that has gone into your pigs, make enquiries to verify that it will match your personal standards, and that the carcase will be butchered and packed to your instructions. Some abattoirs will arrange for sausages to be made and bacon to be cured at additional cost.

You can do the butchering and processing at home, provided the meat is for your own consumption and none is sold. Be aware that the task of dealing with two or more whole carcases is not to be underestimated; there's a great deal of work to be done within a short timescale. We have done it, and it's probably good to tackle this for yourself at least once, to learn what is involved; however we are now fortunate to have a local facility that does it all to a professional standard at a reasonable price.

We prefer to buy and rear four or five weaners at a time. The work involved in looking after five pigs is much the same as for two, but at the end of it there are surplus pigs to sell privately as freezer pork, the income from which offsets the overall cost while the muck heap also benefits. This way, our own year-round supply of pork and bacon costs little or nothing except for the labour involved.

Is it worth it? The answer is a resounding "Yes!", not because of prejudice through personal involvement, but because there isn't any doubt about it: additive-free and humanely reared pork from a traditional breed or cross is beyond any comparison with the insipid, chemically-enhanced mass market equivalent. If you haven't tasted it before, you are in for both a shock and a treat, as the quality is so much higher. We sell our surplus pork at well above the going commercial rate, not for extra profit but because it accurately reflects our cost of production. Despite the premium price, we cannot keep up with demand; customers become regulars, and more approach through recommendation than we can supply. They can't all be imagining the difference.

Above: Early morning portrait of a lamb
Opposite: A small lambing flock

Chapter 7

Sheep

Sheep are arguably the most difficult of farm livestock to keep successfully, so why bother? The challenge of managing them well is rewarded by the range of benefits they offer in return. A small flock can yield fleece, meat, milk and breeding stock, for your own use or to generate income, and all from grassland. Unlike adult pigs and cattle, sheep are of a size and strength that most people can handle without specialist facilities.

You don't have to breed them; some sheep are kept simply as pets, as natural lawn mowers, or solely for their wool. Orphan lambs can be bought in spring for fattening through to slaughter by autumn; but many smallholders want to manage their own flock of breeding ewes.

Which breed should you choose? There is no single, simple answer to that question. It depends on many different factors, not least of which is personal preference; but the advice often given is to buy quality pedigree stock from a reputable prize-winning breeder. However, we strongly suggest that such livestock are not the ideal first purchase for the raw novice, and this applies not just to sheep, but to all livestock.

Pedigree animals have been selectively bred over many generations for specific characteristics that are considered desirable. Inbreeding is commonly used in this process, while relative scarcity often restricts breeding stock to a small number of bloodlines. The promotion of specific attributes may be at the expense of general qualities such as hardiness, prolificacy and longevity. As breeds diverge further from the natural form, there is an associated increase in dependence on human care and intervention to maintain health and well-being. To give a simple example, primitive sheep shed their fleece naturally each spring whereas the more developed breeds need to be sheared.

To establish the reputation of his or her pedigree stock, the breeder needs to show them and to win prizes; but showing promotes physique and visual appeal over utility attributes, so that prize-winning stock may lack the practical characteristics that are of most interest the smallholder. To win consistently around the shows against stiff competition requires considerable effort and commitment alongside expert stockmanship. All this indicates that trying to keep and breed pedigree stock is not for the faint hearted; nor is it for the novice. Instead of starting at the top end of the scale with pedigree livestock, we think the novice is better advised to first gain practical experience with cheaper and hardier stock.

When two pure breeds are crossed, the resulting progeny grow faster to a larger ultimate size than either parent; they are also usually healthier, hardier and more prolific breeders. These characteristics are collectively known as "hybrid vigour".

A flock of cross bred ewes

Commercial farming takes advantage of this in various ways, for example using crossbred ewes to combine the prolificacy of one breed with the heavier lamb carcase of another.

Commercial strains often arise, not from pedigree stock, but from animals with the general characteristics of a breed. The vast majority of such livestock are far removed from any traceable pedigree, but they may nevertheless exhibit the practical attributes of the breed that are of most interest to the farmer and smallholder.

Both crossbred and commercial stock will exhibit a wider variance than pedigree animals, but uniformity may not necessarily be desirable for the smallholder, in fact diversity may be a distinct advantage. Both these types of livestock are easier to source, cheaper to buy, and easier for the novice to keep than pedigree stock. The progression to pedigree stock, if that is the long-term aim, can follow as experience grows.

Another important consideration is tameness. Sheep from the uplands may be as wild as the hills they graze, unused to confinement by fencing and unworkable except with a well-trained dog. Most commercial sheep are worked with dogs and may also fear man by association. However, sheep that are worked without dogs and trained to follow a food bucket instead are easy to manage on a smallholding scale. We think that hand tame, bucket-trained sheep from a small scale breeder are the best foundation for your starter flock.

Bucket trained sheep are easy to manage

Choosing sheep

When choosing sheep, apply commonsense principles as mentioned for pigs, but also examine the teeth, feet, eyes, wool and (in the case of breeding females) the udder. The teeth should be in good condition and bear snugly against the pad of the opposing jaw, not protruding or misshapen.

Teeth indicate age as follows:

- 8 even milk teeth = lamb

- 6 milk teeth plus 2 central larger adult teeth = 1 year plus (two tooth)

- 4 milk teeth plus 4 adult teeth = 2 years plus (four tooth)

- 2 milk teeth plus 6 adult teeth = 3 years plus (six tooth)

- 8 adult teeth = 4 years plus (full mouth)

- Fewer than 8 adult teeth = (broken mouth)

There should be no sign of lameness and the load-bearing surface of hooves should flat without overgrown outer horn, or separation of outer horn from inner sole. The membranes of the eyelid when rolled gently back should be bloodshot, not pale pink, which may indicate anaemia associated with parasites. The fleece should feel greasy with natural lanolin and individual fibres should not display a "break" or weak point along their length that may indicate illness or setback since the last shearing. The udder of a breeding ewe with lambs should have two even-sized teats and feel soft, without any hard areas that may indicate mastitis.

Breeding

Ewe lambs should not be put to the ram in their first year, but grown on to lamb for the first time at two years old. Such ewes with lambs at heel make a good choice for the novice, having proved their mothering qualities, and should bear lambs for several more years to come. As the teeth start to fail at seven or eight years onwards, the ewe becomes "broken mouthed". Once the teeth are worn to the gums or fall out, her breeding days are realistically over as she can no longer support lambs.

Sheep generally find all the sustenance they need through grazing from April to November. Between December and March they require supplementary hay, and it must be good hay. Most sheep are very hardy and can stay out in all weathers, but the land can benefit from resting during part of the winter, and lamb survival rates are higher indoors, so lambing flocks are often housed for several weeks.

The sheep year traditionally begins in the autumn at tupping time. The ram is often described as half the flock and should be chosen carefully. Keeping your own ram can present problems on a few acres because of the need to separate him for most of the year, so hiring a ram when he's needed can be a good option. Make such arrangements well in advance, and ensure that he is wormed and foot trimmed before turning him

A lambing flock needs careful feeding

Above: A ewe receptive to the ram
Opposite: Most lambs are born unaided

out. Ewes served on 5th November will lamb around 1st April.

Ewes require extra food in the latter stages of pregnancy, especially those carrying twins or more, and breeding flocks are fed on a rising plane of nutrition, with bought-in concentrates (organic ewe nuts) introduced eight weeks before lambing. The feed rate steadily increases from half a pound to around two pounds per head per day at birth, continuing for three or four weeks afterwards to support the milk supply. Failure to feed properly can result in pregnancy toxemia, as the developing lambs demand more nutrition than the ewe is consuming, and this condition can be serious, even fatal. Toxoplasmosis, a condition that may cause abortion in ewes, is transmissible to humans, so pregnant women should avoid all contact with in-lamb ewes.

Most ewes will deliver their lambs unaided provided they are given sufficient time. Lambs are correctly presented for birth with nose and forefeet leading together, and intervention is rarely necessary. However malpresentations do occur and these may require assistance. A ewe that strains without result for a long time, or shows the wrong parts of the lamb at her rear end, needs examining to assess the situation. Read widely on the subject, attend a training course beforehand, and ideally have an experienced shepherd on standby to consult if necessary. In a very few cases, veterinary help may be required.

7

It is vital that the lamb suckles the mother's colostrum within the first few hours of life, for this transfers essential immunities as well as nutrition. Most vigorous lambs will achieve this unaided. Weak lambs can be given colostrum by stomach tube, a skill that is useful to learn from another shepherd. After a day or two, when lambs are suckling strongly and have bonded with their mother, the family can be turned out by day and housed at night. After a fortnight they can stay outside round the clock.

Lambs are usually weaned at three to four months old, allowing time for the ewes to regain condition for tupping in the autumn. Spring-born lambs should fatten entirely on good grass to be reach slaughter weight of ninety pounds by late summer or early autumn. However where good grazing is lacking, lambs may need supplementary feeding (creep feed) after weaning to keep them growing. Lambs kept over the winter (stores) become hoggetts in the New Year and many stockmen, including ourselves, prefer their flavour.

Management

Routine management for sheep includes regular foot trimming at intervals of six to eight weeks, unless they have daily access to hard surfaces like stone or concrete to wear down the hoof. Shearing is an annual requirement, although some primitive breeds shed their fleece naturally in the late spring. Small numbers can be shorn using hand shears to avoid the considerable expense of electric clippers, or you might employ a local contractor to do it for you.

Like all stock, sheep should be seen at least daily, and young lambs preferably more often. Learn to know your stock; then anything out of the ordinary is noticeable at once and can be dealt with at an early stage. Some farmers will tell you that a sheep's main aim in life is to die, but in our experience this is not so, and the myth actually arises from poor stockmanship.

Our principles for sheep are the same as for all livestock: start with healthy non-pedigree stock from a small-scale source; inform your choice by common sense; keep them as naturally as possible; avoid overstocking; feed them properly; and learn their normal behaviour through regular observation. Do these simple things, learn through experience, and you should have very few problems.

Many sheep are dosed routinely with chemicals against worms, external parasites and fly strike, and vaccinated annually against a range of diseases; however, much of this can be avoided by strict rotation of grazing once you can operate a closed flock (i.e. no bought-in stock). Beware of liver fluke, a parasite that can kill. The life cycle requires a freshwater snail as an intermediate host, so liver fluke only occurs close to water or on very wet ground. Typical symptoms in sheep are loss of appetite and condition, and standing alone with reluctance to move. A watery swelling can develop beneath the lower jaw while the membranes of the eyelids become pale, indicating anaemia; but heavy infestation can result in sudden death with little or no warning.

Opposite above: Mixed colour triplets to a cross bred ewe
Opposite below right: Foot trimming
Opposite below left: Hand shearing

Our routine is to treat all ewes once a year with a broad spectrum wormer, straight after lambing and before turn out onto clean pasture grazed by cattle the previous year. We dose them again, together with any ewe lambs selected as flock replacements, specifically against fluke in the autumn when the risk from this parasite is highest. Lambs due for slaughter thus remain entirely chemical-free.

Opposite: Wet land carries the risk of liver fluke
Below: Hand sheared and hand spun wool of natural colour

Chapter 8

Cattle

Cattle are much heavier beasts than sheep and so require stronger fencing, gates and handling facilities. I once erected a new post-and-rail fence, using a tractor-mounted post driver for a professional job, only to have it pushed almost flat by five large beef bullocks leaning their combined weight against it to reach the grass on the other side. Cattle will generally respect a single strand of taut barbed wire or electrified fence wire. Standard woven stock wire is not tall enough to prevent them leaning over to flatten it, but extra height can be added with one or two single strands of plain or barbed wire on top, supported on longer stakes, to make an effective fence.

Where ground conditions are soft and wet, cattle cannot be outwintered as their poaching damages pasture, leading to erosion of topsoil and invasion by weed species. Across most of the UK, cattle require outbuildings and yards strong and spacious enough to contain them for five or six months of the year, plus the storage and handling facilities for sufficient hay or silage to feed them.

Opposite: A cross bred bullock
Below: Post and rail fencing

Beef cattle cudding

So why bother with cattle? Cross-grazing with two or more different species of livestock, such as sheep and cattle, has benefits for both pasture and stock. Cattle graze by wrapping the tongue around grass to pull it off, whereas sheep bite close to ground level with their teeth. So sheep can "clean up" after cattle have grazed, utilising the shorter leaf blades that cattle leave behind; and a rule of thumb is that an equal number of sheep will thrive on the same pasture without detriment to the cattle.

Where one livestock species is grazed year after year on the same ground, parasite and disease problems soon build up. The eggs of most internal parasites are shed in the dung and develop into infective larvae that wait on leaves to be eaten. On heavily grazed pasture, the parasite burden can increase rapidly to dangerous levels, especially for young stock like lambs or calves. Chemical wormers are routinely used, with the consequence that resistance to most available treatments is now widespread.

On our smallholding, we use cross grazing with cattle to control the internal parasites of sheep. The available grazing is divided into two halves; one for cattle, and the other for sheep. Stock rotate around the small fields of their allocated half for the whole season; then the following year, they swap over. Most parasites of one species are destroyed when ingested by the other, so by this simple rotation, sheep are always grazing pasture "cleaned" by cattle the previous year, and the parasite burden remains naturally low. There is equal benefit to the cattle.

Red Devon cow with Aberdeen Angus foster calf

Overwintering

We lack the housing and handling facilities for overwintering cattle, so have an arrangement with a neighbouring farm whereby we provide grazing for five or six of their beef steers from May to September each year. This earns a modest rental, usually settled in some form of barter, but the main benefits lie in parasite control as described above, and in conservation grazing of one field by cattle only, as described in the chapter on land management.

Friends kept a house cow for some years on a three acre smallholding. The main field was divided into two paddocks, with one cut for hay while the other was grazed. Twice a day she was led across the road to an old stone barn for milking; then she was housed in the same barn for the winter, fed mainly on the hay.

A drawback of keeping one cow is that she needs to calve every year and for that must "dry off", so there is a discontinuity in the milk supply. For year-round milking, two cows are needed, calving at different times. The volume of milk given (three or four gallons daily per cow) far exceeds the liquid requirement of any family household, so the surplus must be processed into dairy products. This, plus the twice daily milking, makes a house cow the biggest livestock tie of all, and one not be taken lightly.

Red Devon suckler herd

Beef cattle

by Pammy & Ritchie Riggs

Cattle are herd creatures so must be kept with at least one other of the species. Common practice is to keep either a small suckler herd of cows that produce and raise their own calves; or to grow on a number of young cattle to beef weight. Buying in very young stock is a specialist area.

The indigenous breeds are reputed to give superior beef and are suited to their region: Hereford, Red Devon, South Devon, Red Sussex to name only a few. The Scottish Aberdeen Angus and Highlands are also well suited to smallholdings, some have horns and some are naturally polled (without horns). They are generally smaller and more docile than the commercial continental breeds. The Dexter is a very small breed whose history is linked closely to smallholding. For first time buyers it is advisable to visit breeders and take good advice, or to buy from other smallholders. Leave cattle markets to the more experienced herdsman.

You can buy young stock ready to fatten to beef weight or buy weaned calves at six to nine months old. These will have had a good start in life and should be able to live outdoors most of the time. A small suckler herd will take more expertise, not least in

A large calf suckling

registering for cattle passports and ear tagging, but there are rewards in rearing your own calves from birth. The cow will need either to visit a bull or be artificially inseminated (AI). There are many companies who perform this task for a fee, and by arrangement you can obtain semen from any breed. Think carefully what would be best for your situation and choose breeds the same as, or smaller than, your cow for easy calving. AI is not as efficient as a bull, and a degree of close observation and record keeping is required to tell you when your cow is "bulling". Transporting your cow to a bull can be expensive and you need to follow the regulations for animal movements. If TB restrictions are put in place on the hosting farm your cow can become stranded until the all clear is given.The gestation period for cattle is nine months; it is best to calve down in the spring or early summer, when the cattle are outdoors and a good flush of grass will ensure enough milk for the calf.

A beef animal is ready for slaughter when it is "well covered", with a degree of fat. A rough guide would be 450 kg live weight for a smallish native breed; this will yield about 250 kg of usable meat. There is a cut-off age of 30 months when some abattoirs will not be able to handle your animal. Find out in advance if your local abattoir has a service for preparing and cutting your beef carcass and take advice from them. Regulations change remarkably often and the abattoir will be up to date.

Strip grazing a field by moving an electric fence can be a very efficient way to utilise smaller areas. Remember to provide clean drinking water at all times. On a smallholding, preserving your pasture is very important. Keep to a low stocking density of one beast per acre until you understand the limits of your land. Rotation of pasture helps to keep the worm burden to a minimum, but regular worm counts by your vet will give an accurate picture and enable any problems to be addressed.

It is normal practice to house cattle over winter and feed them hay or silage. Buying in feed is expensive, and there is the problem of handling heavy bales, since very little small bale hay or silage is made nowadays. Winter housing should be airy, suitably sheltered from driving rain and with one part well bedded down with straw or sand. This can be deep litter or selectively mucked out. Cattle need a constant source of clean drinking water, and drink a great deal when they are eating dry food. A ring feeder or robust hay manger is essential, as forage that has been fouled will not be eaten.

One vital piece of kit is the cattle crush, a restraining structure necessary for holding a large beast still. There will be visits from the vet, at least for TB testing. Walk the cattle through the crush from time to time as a training exercise, it will make life much easier on the days when it is needed.

Strip grazing Red Devon cattle using an electric fence

A cattle crush

Above: Inspecting a top bar hive
Opposite: Examining comb from a top bar hive

Chapter 9

Other livestock: bees, rabbits, goats

Bees

by Phil Chandler

If you grow top fruit, beans, almonds, coppiced hazel or willow, flowering crops of any kind, or just have plenty of wild flowers, you will already have bees visiting, so keeping a hive or two of your own would seem a logical development.

Compared to most livestock, honeybees need little attention and so can be added to the smallholding without fear of creating a serious drain on your time. However, as with any other creature that comes within our care, they do have needs to be addressed, and someone must give them the right kind of attention at the right times. Honeybees are – and will always remain – wild creatures, unimpressed by any attempt to domesticate them. "Keeping" them, therefore, is really a matter of providing them with suitable accommodation and allowing them free rein to roam. Beyond that, you have to consider the degree and style of management you will endeavour to apply.

A small apiary of Dadant hives

The chances are that flowering crops you grow are already being pollinated quite effectively by wild bees and other insects; and unless you grow such crops on a large scale, adding honeybees to the mix will have only a marginal effect on yields. Exceptions to this might include areas where neighbours routinely spray with insecticides, with the result that wild insect numbers have been drastically reduced; or places where wild bee populations have suffered for other reasons, such as heavy pollution. Unfortunately, in either of these cases, you are probably in the wrong place to keep honeybees. Honey yields are dependent on three main factors: the number of colonies kept, the extent and variety of available food, and the weather. Of these, only the first is fully under your control, as bees will forage over a three mile radius from their hive. If most of that territory is flower-rich meadows and hedgerows, or verdant, uncultivated wilderness, you are probably well placed to keep at least half a dozen hives if you so choose. In a good year you might reasonably expect between 25 and 75 kilos of honey per hive. Regardless of other factors, in a wet, cool summer, you may not be able to harvest more than a kilo or two.

A fundamental choice you have to make is between "conventional" beekeeping, using variations of the Langstroth frames-and-foundation hive; and "natural" beekeeping, which is mostly based on variants of the top bar hive. The route you follow will depend on your philosophy, your priorities and your pocket. The conventional approach requires a substantial initial investment in equipment, an ongoing dependence on bought-in supplies and the possibility of higher yields; while the natural path can be followed at a minimal cost, with generally lower but more sustainable yields. The horizontal top bar hive is best suited to people who are able to give bees a little time fairly often - especially in spring and summer – while vertical hives, such as the Warré and the Perone, are designed to be left alone throughout the season, although there is always a risk of losing bees from swarming. Before choosing between them, you should first seek out opportunities to have some direct, hands-on encounters with live honeybees en masse. Not everyone is temperamentally suited to working with bees, and it is as well to establish this one way or the other before you find yourself with tens of thousands of them in your back yard.

Given the opportunity to sell natural, in-the-comb honey from hives free of synthetic medications as a premium product, natural beekeeping will appeal to many smallholders, and if you are among them, you need to invest some time to learn about bees and beekeeping from this perspective. There are beekeepers' associations in most urban and many rural areas that run open days and courses for beginners, and although they are mostly orientated towards the conventional approach, some of them are starting to accommodate more natural practices and are increasingly likely to have members with top bar hives.

A typical year with top bar hives might look like this:

- Late winter/early spring – open one end briefly to check the bees have enough food. If necessary, provide fondant. When warm enough, check for mites and treat colony with powdered sugar if necessary.

- Spring/early summer – check the queen has room to lay and that pollen and nectar are coming in. Swarm management may be necessary if the colony is building strongly.

- Summer – routine checks for signs of swarm preparations (full examinations rarely necessary). Some honey may be harvested if nectar is plentiful and it is likely to be replaced.

- Late summer – more honey can be harvested if plentiful, leaving sufficient for the bees to over-winter. Check for mites and treat colony with powdered sugar if necessary.

- Autumn – check bees winter stores and feed if necessary. Add insulation on top of hives and ensure some ventilation at floor level only, to avoid condensation.

- Winter – read books about bees!

Beekeeping is a fascinating and absorbing activity that has the potential to enrich your relationship with the landscape and its untamed inhabitants.

New Zealand white rabbits six weeks old

Rabbits for meat

by Alison Wilson

Farmed rabbit meat is much more delicate than the gamier taste of wild rabbit. New Zealand Whites or Blacks are ideal meat rabbits, as are Californians. These will all give a good amount of meat as well as high quality fur. The pelts can be tanned at home or professionally and then used for making garments, or glove and boot liners.

To keep a doe and a buck you will need a separate hutch for each, plus larger accommodation for the youngsters between weaning at six weeks and the table at 12 weeks onwards. That is a minimum of three hutches. Meat rabbits need larger than average accommodation, especially for the doe who could have up to 10 kittens with her for six weeks. Hutches based on an eight by four feet sheet of ply wood constructed with three sections, two with wire mesh fronts and one solid one will be the right size for the family of youngsters. Hutches half the size, with two sections, will accommodate the adults individually. Hutches can be positioned inside or out, but need to be in the coolest place possible. Rabbits suffer much more from heat than they do from cold.

Food can be given in a bowl or hopper, but these must be ceramic or metal as rabbits will chew anything else. Normal rabbit bottle drinkers are fine, but meat rabbits drink a great deal, so buy the largest ones that fit on the cage front. Have enough

A double bank of hutches

bottles on the cage front for the occupants to drink without having to share. Have a spare set full of water so that cages can be fed and watered quickly, refilling empties later for the next day. Keep bottles in a container for carrying as if you drop a full one, the spout is liable to snap off.

Complete pellets are the easiest way of ensuring rabbits are kept in prime condition; with muesli style food there is too much margin for them to leave the nutritious bits. The pellets should contain at least 20% protein to maintain breeding condition for adults and steady growth for their young. An agricultural feed merchant will have a better supply than a pet shop where the rabbit food is intended for pets, not meat rabbits.Hay should also be provided in a rack to keep it off the floor, as it helps to provide the fibre needed in the diet. It also reduces tooth growth due to the chewing action needed to break it down. Food should be stored in a vermin proof container kept near the rabbitry. With a set of full water bottles available, feeding the rabbits once a day should then be a fast process.

Breeding:

- A doe is sexually mature at 20 weeks and a buck at 24 weeks.

- A doe is pregnant for 30/33 days.

The doe will come into season immediately upon being mated, so at the right age put her in with the buck. Introduce her in the morning, and again the same evening. This will ensure a bigger litter then would have been achieved on a single mating. Supervise the encounters as the doe can become aggressive or the buck may not be able to catch her.

A few days before she is due to kindle (give birth) move a doe that has been sharing to solitary accommodation. Provide her with hay to make a nest, and if there is not a separate compartment in her hutch, provide a plastic box to help her keep the nest together.

Keep records of each mating and the number of kittens, to ensure that you provide for the doe in time for her to kindle, and know when the youngsters are the right age for the table. Breeding records are kept through the male line, so if you plan to expand the rabbitry beyond the initial buck and doe you will need another unrelated buck. Do not interbreed as deformities will occur very quickly.

Start up check list:

- Housing

- Location

- Bowls/hopper feeders

- Drinker bottles

- Carrier for bottles

- Bottle brush

- Hay mangers

- Rat proof food storage

Anglo Nubian milking goat

Goats

by Penny Hall

Goats are sociable creatures and given adequate feeding, housing and commitment make a useful addition to the smallholding.

Tensioned stock fencing with two strands of electrified wire on the inside will keep most goats contained. The electric fence must be in full working order and tested regularly. Goats caught up in it, even if it is not working, have little chance of survival. If it is working they simply do not go near it. Tethering is not an option; refer to current goat welfare regulations.

Shelter from sun, wind and rain is essential. A three sided construction in the field for daytime, and a draught-proof loose box with a bench to lie on for the night, suit their requirements. Straw for bedding can be deep littered during the winter, not only saving labour but providing the goat with good insulation at floor level.

Unlike sheep, goats are browsers rather than grazers. They will, however, graze off the long grass and some weeds which sheep and horses find unpalatable. Good quality hay, straw and concentrates with the addition of leafy branches cut from the hedgerow and excess from the vegetable garden will keep the goat well nourished. Water must be

Anglo Nubian kids after separation

available at all times and changed several times a day, with buckets kept clean and uncontaminated. Goats prefer warm water especially during the winter.

The Swiss types of goat (Saanen, Toggenburg and Alpine) are all excellent milkers. It is not unusual for them to produce over a gallon of milk a day and need only to be kidded on alternate years; some will continue milking for many years without kidding. Anglo Nubian goats, however, need to kid each year; their yield is slightly lower but the butter fat content is higher, so this milk is considered superior for making cheese and yoghurt. The Angora Goat produces a luxurious fibre but only enough milk to feed her kids. It is not a good idea to buy your first goat from a market where animals are often of unknown parentage, small, undernourished, and unlikely to make a good dairy goat. Better to visit an agricultural show where goats are being exhibited, look at the different breeds and talk to the exhibitors, who are mostly very helpful and willing to pass on their knowledge to beginners. They will also have goats for sale at home, from milk recorded stock. They will advise you to buy two, probably of different ages.

Do not even consider buying a male goat. Your local goat club will have a list of male goats at stud.When breeding to the next generation of milking goat, use a male of the same breed. If rearing for meat, a Boer goat sire produces quick maturing kids ready for slaughter at six months.

The gestation period for a goat is 150 days. Once kids are born, arrangements must be made with the vet to disbud them and castrate the male kids; this is usually done within a few days of birth. It is not advisable to let any dairy goat grow horns, as they can be dangerous to other goats and humans.

Sad as it may seem, disbudding time is the best time for the kids to be taken away from their mother. Both parties adapt very quickly, the kids depending on bottle feeding for several months, the mother becoming your milking goat providing enough milk for the house, the kids and a bit left over for making cheese or yoghurt. However she will need adequate and careful feeding of concentrates and fibre during this time if she is to produce the high yields of milk expected.

A decision also has to be made regarding the male kids. The best thing is to raise them for meat, otherwise they risk a miserable life as a pet that nobody cares about.

Goats are creatures of habit and thrive best when each day follows the same routine. Leftover food should be removed, buckets cleaned and water changed at least twice a day. Hay should be replenished in the hayrack; any hay dropped on the floor will never be eaten, so use it for bedding. Bedding should be forked over daily and fresh straw put on top if the deep litter system is to be used.

Hoof trimming must be done every six weeks, for if this is neglected and hooves grow too long they become misshapen. Young goats with overgrown hooves often develop arthritis later in life from walking awkwardly to avoid pain. Overgrown hooves also develop foot rot very quickly. Be advised by your vet on vaccination and worming, as guidelines change all the time.

British Saanen goat due to kid

Chapter 10

Land management

Stock proof field boundaries are a basic prerequisite for most land management. Where stock is not intended to graze, the boundary keeps animals out; while on grazing land, the boundary contains animals where you intend them to be. Pasture is usually sub-divided into smaller paddocks that enable sound principles of rotation to be followed.

Fencing is expensive to buy and time-consuming to erect, so requires careful planning. Gates allow movement from one enclosure to another, and require substantial hanging and latching posts set well into the ground – more expense!

Ideally, each livestock species requires a different style of fencing, but for sheep and cattle, the usual solution is stock wire mesh supported by wooden stakes, drawn tight between straining posts and topped with a strand or two of plain or barbed wire. Post and rail fencing is more expensive but popular for horses, works for adult cattle, and can also support wire mesh against sheep and calves if required. A single strand of tight barbed wire at the correct height will hold contented cattle, although they may break through it if hungry or frightened.

Look carefully at fencing erected by contractors in the locality for guidance on design and construction. The principle for wire fencing is to work in straight lines between straining posts set deep into the ground, and firmly cross-braced to resist movement. Intermediate stakes support the wire vertically but carry no strain. Wire is fixed firmly to straining posts but allowed movement at intermediate stakes by driving staples to half depth. Posts can be dug in and stakes driven by hand, but best results are obtained with a tractor-mounted post driver. Wire is strained with a purpose-made device, by careful use of tractor power, or simply by fixing each end before pulling back at changes in direction.

Electric fencing can be a semi-permanent fixture, for example a woven mesh fence surrounding free range poultry, as much to keep foxes out as to keep the birds in. Alternatively it can be temporary, for rolling up and moving to a fresh location with relative ease. Energisers powered by mains or rechargeable battery are initially more expensive to buy than smaller consumable battery types, but are much cheaper to operate in the long run, and give a more powerful deterrent shock into the bargain

Opposite: A layed hedge

Landscape features

Stock proof boundaries can also take the more permanent form of landscape features such as stone walls, turf banks or water-filled dykes. These obviously involve more time and labour to construct, but may cost little or nothing where local materials are used; and once in place can last for centuries with a little maintenance.

On our smallholding, we have earth banks with hedges on top, sometimes faced with stone at vulnerable areas such as gateways. These banks require a batter, or slope away from the vertical, of around one-in-four when built or repaired, to avoid the risk of collapse as the bank settles and spreads outward over time. Stonework starts with a trench dug down to firm subsoil, then a foundation course of the largest stones, followed by courses of decreasing size with increase in height. Joints are staggered, as in brickwork, to avoid weak vertical faults. Dry earth is infilled firmly behind, and the correct batter maintained, as work proceeds.

Turf-faced banks are built or repaired in similar fashion, starting with a foundation trench and staggering the joints in successive courses. Turves are dug from the field as required and infilled behind with rammed earth.

Above: A tractor mounted post driver
Opposite top left: Correct batter for stonework
Opposite top right: Turf facing an earth bank
Opposite below: Stone facing an earth bank

Hedge laying

We are steadily removing many of the wire fences that were put up on our smallholding when we first arrived. Back then, fences were necessary, but allowing neglected hedges to grow tall, and then laying them on a regular cycle, has largely eliminated the need for fencing whilst enabling us to become self-sufficient in firewood.

The basis of hedge laying is to make a sloping cut near the base of a woody stem, slicing about three quarters of the way through to leave a hinge of living material. The stem can then be carefully bent over towards the horizontal and woven among stakes, or otherwise secured in place. Surplus stems are cut out as work proceeds. New shoots arise from the cut stumps, while layed stems continue to live and grow via the hinge of sapwood. The combination of vertical regrowth plus angled layed stems produces an impenetrable woven living barrier. Hawthorn and blackthorn are most effective against livestock, but all native hedge species can be managed in this way.

Layed hedges provide valuable shelter from the elements to crops and livestock, while selected young trees within the hedge line can be left to grow on. Hedges managed in this way yield a regular harvest of firewood, garden sticks and greenwood for light construction, and eventually timber from hedgerow trees, while providing valuable food and shelter for wildlife. In addition, a layed hedge simply looks so much better in the landscape than a flailed hedge or fence. Laying a section each year in rotation creates a mosaic of regrowth across the smallholding from first year shoots to mature hedgerows, providing maximum benefit to wildlife.

Opposite: Re-laying a gappy hedge to make it stock proof
Below: Luxuriant regrowth following hedge laying

Mowing hay

Grassland

Once grazing livestock can be contained, grassland can be managed. Grass leaf growth accelerates to a maximum in late spring, slows right down during seed formation in early summer, then increases modestly through the autumn until winter temperatures cause growth to cease.

Grazing management is a delicate balancing act between the available fodder and the number of mouths eating it. The ideal is to stock a field heavily in the spring and graze down to one or two inches of leaf remaining. Move stock on to rest the pasture

There are several controls that can help you to achieve the right balance. The number of mouths available naturally increases during the spring, when most lambs and calves are born, and decreases towards autumn as fat stock are sent for slaughter. More stock can be brought in temporarily for part, or all of the grazing season. Excess fodder can be allowed to grow on and conserved for the winter as hay or silage. Topping with a mechanical cutter drawn behind a tractor or quad bike cuts flowering stalks to encourage leaf production. Applying fertiliser boosts fodder production.

As explained in previous chapters, our aim is to graze half the land with sheep, and half with cattle, setting aside perhaps two acres for hay. Each year, the sheep swap over to follow the cattle of the previous season. Most parasites of one are killed through ingestion by the other, so the parasite burden of the pasture, and reliance on chemical wormers, is greatly reduced.

Baling hay

Haymaking on a smallholding scale involves "shutting up" one or more fields before spring growth starts, and mowing as the grass flowers, but before seed is set, for maximum protein content in the leaf. You need four consecutive hot, sunny days without rain to dry the crop sufficiently. The crop needs turning three or more times to dry evenly. When hay is dry enough to crackle between your fingers or the hay turner but retains a hint of green, it's fit to bale. If weather permits, stook the bales on the field to air for another day or two before carting to dry store.

Neighbours or contractors with appropriate machinery can be called upon to do all this for you, but the problem with this approach is that your small acreage is likely to be low on their priority list, so windows of opportunity can be missed. Our choice is to make hay with our own machinery, rather than rely on someone else. Of course it is possible to make hay by hand or horse power as well. However you tackle it, hay making is hard, hot, dusty work with no guarantee of success at the end of it, but there is great satisfaction once your own hay is safely stacked for the winter ahead.

Having taken a hay crop, you must put something back to replace the nutrients removed from the soil. Muck spread after cutting is a good way to achieve this. The regrowth is relatively clean of parasites and so is ideal for grazing by weaned lambs or other young stock.

Drainage

Most agricultural land in the UK is drained in some way, to lower the water table and increase productivity. Wet land is cold land, slow to start growing grass in the spring and early to cease in the autumn. The first principle of drainage is to intercept surface water running downhill before it flows into the field; hence open ditches often run along the upper boundary of a field to collect and carry away water down side ditches. There may also be a network of pipe drains below the field surface, feeding into the side ditches, to capture and carry away direct rainfall or groundwater.

In general, any ditch holding standing water isn't working properly and needs investigation and clearance. Expect all ditches to require cleaning out every few years to remove accumulated debris, otherwise they silt up to become ineffective. A waterlogged patch in a field presents a hazard to man, beast and tractor, so dig down to clear a blocked pipe drain, or put in a new one to solve the problem.

However, the smallholder can also employ diverse ground conditions to advantage. Wet land may provide lush grazing in a hot summer when the drier slopes have all scorched brown. On the other hand, these same dry slopes grow an early bite of spring grass ahead of wetter ground. In our area wet, rushy pasture is often drained, ploughed and reseeded for higher productivity, but instead we make use of the soft rush that naturally grows here by cutting, drying, baling and storing it for winter bedding, reducing costs and reliance on bought-in substitutes like straw.

Cleaning silt from a ditch

Ragged robin thrives in wet meadows

A pond increases biodiversity

Wildlife

Tightly grazed pasture is not the ideal habitat for most wildlife, so as stewards of our land, we all have a responsibility to manage some areas of the smallholding in a different way, and where possible to create new habitats for maximum diversity. Avoiding the use of chemical fertilisers or weed killers is a good first step. A pond, of any size, is an extremely valuable addition, especially with a patch of wetland surrounding it. Log piles or brash can be left unburned after hedge laying or tree felling, and dead trees left standing or fallen. Native trees can be planted in awkward field corners or odd patches of ground; or better still, plant a wood. Another approach is to simply fence out grazing stock and allow colonisation by trees and shrubs.

Saplings or cuttings are best planted during winter dormancy. Fast growing species like willow, ash or poplar soon impact upon the landscape, reaching heights of 20 feet or more within a few years. Most native species respond well to coppicing, sending out several new shoots when the stem is cut down close to ground level. Small areas can be harvested on a regular cycle for fencing, structural work or firewood, becoming steadily more productive over time while creating a range of habitats for wildlife to exploit. Smallholders planting an acre or two of fast growing coppice can become self sufficient in wood fuel within a few years.

Female broad bodied chaser in wet grassland

Above: Willow catkins are an important food source for wild bees
Opposite: Coppicing in progress

Grazing and cutting regimes of grassland can be changed to accommodate existing wildlife or to encourage colonisation by new species. For example our wettest field, The Marsh, is grazed by cattle only - never with sheep. The difference is obvious as you walk into the Marsh; it's colourful with a succession of flowers throughout the spring and early summer, and alive with the various creatures whose life cycle includes these plants. We've created a pond here by flooding a drainage ditch along the lower boundary, to attract more dragonflies, amphibians and birds.

A further two-and-a-half acres of our smallholding are taken permanently out of grazing use. Here we have created wildlife ponds and wetlands; planted trees and coppiced some; encouraged tussocky rough grassland; and carried out other environmental management. This could be seen as non-productive land, yet our experience has been that, alongside the obvious benefit for wildlife and landscape, such stewardship can also bring financial rewards in the long term, linked to environmental funding or tourism.

Chapter 11

Tools and machinery

Once your smallholding activities extend beyond the scale of a garden plus allotment, you soon find the need for specialised tools and machinery that empower your work, and without which you would find it difficult or even impossible to cope.

Edge tools

This is a broad category of tools made from hardened and tempered steel, with the common denominator of a bevelled and sharpened cutting edge.

Bill hook, grass hook and staff hook are all variations on a theme. Bill hooks are short handled blades designed to cut green wood of small diameter for hedging, coppicing, fencing, hurdle making and other associated crafts. The curved edge gathers twiggy material together and guides the blade to bite in rather than slip away from the cut, while the hooked nose is used for leverage when laying or splitting stems. Grass hooks are short handled, deeply curved blades designed to gather and cut grass or other light-stemmed plants. The staff hook is simply a long handled variation of the grass hook.

In the traditional manufacturing process, red hot steel is forged to give shape, strength and balance. Mass produced tools are cold pressed from steel of uniform thickness, a process that can never reproduce the tapered cross section of a forged blade, so my strong preference is to use forged tools. Look for the only remaining UK manufacturer, Morris, or keep an eye open for quality second hand items.

The key to personal choice of bill hook is the feel of the tool in the hand. Weight and sharpness, rather than muscle power, drive the cut through wood, so a heavier bill hook can actually prove less tiring in use over a long working day. Conversely a lighter blade with more "hook" to the nose is easier to use for splitting hazel rods. Match the tool to the task for the best results, or choose a general purpose type that will do most jobs reasonably well.

The same applies to the choice of grass and staff hooks, which benefit from a certain weight and balance to cut sweetly with a minimum of effort. The longer arc of swing of the staff hook makes lighter work of tough customers such as brambles or blackthorn suckers, and makes it possible to cut right in to the base of an overgrown hedge. Still, the grass hook has a place for trimming in tight corners or any restricted area where the long handled staff hook cannot be swung in safety.

Opposite above left:: Sharpening a bill hook
Opposite above right: Hand tools for the smallholding
Opposite below: A selection of loppers

Above: Hand tools for hedge laying
Opposite: Splitting logs with an axe

Axes are used for specific tasks, so the weight of the head needs to match the work being done. A seven pound axe on a long handle delivers a powerful blow to split large logs into sections of firewood, while a lighter axe of one or two pounds on a short handle is ideal for splitting lighter sections into kindling. Traditionalists might also use an axe for felling or laying large diameter stems whilst hedging, for chopping stakes to length, or for cutting older coppice wood such as sweet chestnut poles.

All edge tools are safer and less tiring to use when sharp. A keen blade will bite into the wood whereas a blunt edge may skid away, so it's important to sharpen these tools when necessary and avoid damage to the cutting edge in use. Sharpness is best maintained by means of a flat whetstone, held against the bevelled face and moved in a steady circular motion until the cutting edge is razor sharp. The curved edges of grass and staff hooks are sharpened in similar fashion with a round whetstone.

Saws and loppers

A bow saw with a two foot blade is useful when hedge laying for cutting or trimming middle sized green wood that is too large for the bill hook but on the small side for the chainsaw. Use a "wet wood" blade with raker teeth on greenwood for easier cutting. A compact pruning saw with a folding blade slips easily into the pocket and is often handy where space is too restricted for larger saws to be used, for example

Above: Tools for digging and tamping
Opposite left: Using the Devon shovel
Opposite right: Tamping with an iron bar

when cutting out surplus rods from a tight stand of hazel to leave those selected for laying. A stout pair of loppers with extending handles are also useful. Modern hard-tipped saw blades hold their sharp edge for a very long time if used correctly, that is, only on green wood and avoiding any contact with earth or stone.

Digging and tamping tools

The Devon shovel appears a strange shape to the uninitiated, with its cranked, heart-shaped blade mounted onto a very long plain handle. However, when you first attempt to dig a post hole three feet deep, you soon discover its advantages. The cranked blade enables earth to be scooped and lifted clear of the hole, the pointed shape of the blade reaches into the corners, while the long handle remains effective down to full depth. For general use in building or field work, the long handle enables you to work in an upright posture, reducing strain on the back.

In hard or stony ground, spade or shovel may be unable to dig, and here you need a chisel pointed iron bar. The bar is matched to the task in hand and the stature of the user. To dig a post hole for example, the bar is held vertically and repeatedly plunged down into the base of the hole, using its weight to break through and lever sideways stone or hard packed soil. The Devon shovel is brought into play to scoop out the loosened spoil, then the bar used again alternately, until the required depth is reached.

To set a post in the hole, a little spoil is shoveled back in and the iron bar reversed, so that the flat head tamps it down hard. More earth is shoveled in and tamped down alternately until the hole is refilled. If the tamping is done correctly, all the spoil will fit back in and the post will be firm.

The drivall, rubber maul or sledge hammer can all be used for driving fencing stakes into the ground; but the sledge hammer is the most versatile of the three. Apart from many other uses in building and general work, the sledge hammer can drive in fencing stakes with minimal damage, provided the top is struck squarely by the hammer face. Alternatively, make up a sacrificial strike plate from wood or hard plastic to protect the tops against damage.

Earth banks become damaged over time by the elements, by burrowing animals and by the feet of browsing livestock. Typically the base slumps outwards, and is repaired by "footing out" the accumulated spoil and "casting up" onto the top of the bank. The traditional tool for footing out is the mattock, a narrow, slightly curved digging blade set at right angles to a handle of medium length. A downward swing cuts into the earth, using the weight and momentum of the head rather than muscle power. The Devon shovel is then used to scoop up the loosened spoil and cast this up.

The mattock is also handy for the shallow cultivation of garden soil when, for example, a seed bed is being prepared from ground compacted over the winter period. Swinging the mattock in a short arc with a chopping motion, at a shallow angle to the ground, breaks up the hard surface in a controlled manner, leaving the soil to dry out for raking into a suitable tilth for sowing.

Power tools

There's no denying the usefulness of the petrol-engined strimmer/brushcutter, though some people today find their noise, vibration and pollution too much to bear and are reverting to scythes instead. Where there is a need to manage tussocky rough grasses, rushes, nettles, thistles, brambles and woody scrub, the strimmer offers faster progress in return for less effort than hand tools. It also brings some tasks within reach that would otherwise prove impossible; for example, to clear vegetation from a chicken wire fence without damaging the mesh.

For heavier work such as cutting rushes or brambles, a steel cutting blade is substituted for nylon cord. The low budget models are not powerful enough for the rigours of brush cutting, so choose a mid-range machine designed for heavier work, ideally with a shaft drive to the work head, rather than cable. When buying, think ahead to future maintenance and choose a quality brand from a local supplier that can provide parts and service.

The chainsaw similarly saves time and effort compared to hand tools, and is especially useful in confined spaces such as the baseline of a dense hedge or a thicket of poles where axe or billhook cannot be swung. Electric saws are lightweight, quiet and clean in operation, but are limited to logging firewood close to a mains supply. A petrol engined model is necessary for all fieldwork, being slightly heavier, noisier and smellier, but portable anywhere; choose a mid-range model of a quality brand.

Like all power tools, chainsaws are potentially dangerous and should be handled with respect. Appropriate safety clothing should be worn. A well maintained saw with a sharp chain is safer to use, so read the instruction booklet to learn the maintenance and sharpening procedures,. To use a chainsaw on land other than your own, the law requires that you hold a valid certificate of competence and refresh your training at intervals.

Logging firewood with a chainsaw

Tractor-mounted disc mower

Tractor

Our tractor and implements date from the 1960's and 70's. These are too small for the modern commercial farmer; not quite old enough to attract high vintage prices; and crucially contain no plastics or electronics that are fiendishly difficult to repair. Metal may be rusty, bent, worn or broken - but it can usually be reshaped and bolted, welded or riveted together again. Machinery from this era is therefore cheap to buy and straightforward to maintain. Be aware that smaller tractors predating this era may lack the power required for some tasks, for example to operate a baler on a sloping hayfield.

Our Massey Ferguson dates from 1964. The paintwork is faded, the windscreen is cracked, most of the electrics don't work – but though we don't care much for appearances, we do carry out essential maintenance so that it continues to deliver what we ask of it around the smallholding. If you're not mechanically minded, seek advice from a local agricultural service and repair workshop when considering a second hand tractor. Pay them to inspect the one you fancy before purchasing - that should protect you against making an expensive mistake.

Turning hay

Implements

We used to cut for hay with an old finger-beam mower, but this had shortcomings that caused frustration and wastage, so eventually we replaced it with a disc mower, which does a much better job. Basically it has four rotary cutting heads mounted on a flat beam that slides on the ground, gliding under the crop being mown to leave a flat swathe behind. The cutting blades are easily resharpened or replaced. All in all, it's solidly made and has given no problems at all since purchased second hand at a farm sale more than a decade ago.

Turning hay by hand under the hot sun is often portrayed as an idyllic rural occupation - by those who have never done it! It's actually very hard, hot, dusty work, so you can see why implements were designed to mechanise the operation. The popular haybob style of turner has side gates that can be set to deflect the thrown hay to right or left of the tractor, enabling you to clear the outside swathe from the hedge or fence line for example; or to "row up" ready for baling.

The small square baler is the most complex implement that most smallholders will need to use. Balers were ruggedly made to give many years of service, and some are still available on the second hand market today. As with the tractor itself, if you lack

Above: Link or transport box
Opposite above: Baling hay
Opposite below: Carting hay

the confidence to make a decision, seek professional advice before you buy. Our current model was found rusting on a neighbouring farm, awaiting disposal as scrap for want of a critical part. I had salvaged that part from a previous baler, so I soon had a working baler again at very low cost. Balers are especially vulnerable to rust damage so store them carefully out of season, coating the vulnerable parts with oil and keeping in a dry outbuilding, or covering the whole with a tarpaulin against the weather. Lubricate moving parts carefully before use.

Hay bales require a sizeable trailer to handle any quantity. Our choice is a low-loader, having the floor slung at axle height rather than above the wheels, to reduce the height that bales must be lifted. It measures 12 feet by 5, with lades at each end to support a tall load, and will comfortably hold 60 bales. Two or three trips with this will clear one of our small fields in an evening. We bought it at a farm sale over 25 years ago, have reboarded the floor and had some welding repairs carried out to the chassis. It stands out in the weather all year, so has lasted well in the circumstances. It also comes in handy for carting other bulky loads.

The link or transport box is arguably the most useful single piece of kit to complement any tractor and mounts directly onto the three-point linkage. One face of the box hinges aside or can be lifted clear, making it possible to walk in livestock like sheep or pigs, or to load things that are too heavy to lift manually. It's so versatile that we use it for all manner of tasks around the smallholding, wherever anything has to be moved from one place to another. Ours came with the tractor, and has recently been overhauled for the first time in 25 years, with replacement iron members welded around the floor and new exterior grade plywood boards fitted.

Livestock trailer

Any smallholder keeping livestock soon needs a trailer to move them from one place to another. Trailers come in various forms depending on their intended load, and many are too large for a family car to tow, needing a heavier and more powerful four-wheel-drive vehicle. We only transport small numbers of sheep or pigs, so manage with a small car trailer measuring six feet by four. It's homemade, though not by me - it came from another farm sale. There's a steel chassis clad with wooden boards and a sheet steel roof. Today, look for a trailer made of aluminium rather than wood, as the modern obsession with disinfection has already made the use of wooden trailers unacceptable at some shows, and this prejudice could become widespread in the future.

You will have realised by now that there's no need to spend a fortune on the latest high-tech equipment, for most can be bought second hand at farm sales or through local contacts at relatively low cost.

Safety

A final word on this important issue. This chapter presents an overview of tools and machinery used on our smallholding, but it cannot be held responsible for any accident or injury that readers may experience. The safe use of tools is about more than just wearing protective equipment; it is the user's responsibility to follow safe working practice at all times. Think to ensure a safe working zone before you start. Communicate effectively with fellow workers or bystanders so that they know what you expect of them.

Keep in mind that all tools are potentially dangerous. A wise precaution is to attend an emergency first aid course and then subsequent refresher courses at intervals. Carry a pocket-sized first aid kit of bandages and wound dressings when using cutting tools at any distance from the house, and tell someone where you are working and when you plan to return. Carry a mobile phone for summoning help fast if necessary. Know your OS map reference, and keep it with your phone for accurate location by the emergency services; this is especially important in rural areas lacking street names and house numbers.

Take care, work safely and enjoy your smallholding.

Moving pigs with a small car trailer

BORLINN VALLEY

RL-01B2-EU

ORGANIC FARM

I.O.F.G.A. (811) AND

BASKETRY

FROZEN MEATS

Eggs, VEGETABLES,
TREES, Fruit,
BASKETS MADE TO ORDER.
SUPPLIERS OF ORGANIC
Animal Feed

Cheaper than So called "Free R

ORGANIC
CHICKEN
€9.00/kg

lettuce €1.60 per hea
Spinach €5.00/kg

DuckLing €

Chapter 12

Finance

"Is there a living to be made in smallholding?" This is the key question that potential smallholders usually ask, but it has no easy answer. Certainly, the unplanned and haphazard sale of vegetables, eggs or meat that are surplus to the needs of a self-sufficient household will not generate much income. It is a fact that many commercial farmers, with large acreages of land at their disposal, are unable to earn a living solely from farming activities, and rely on supplementary income from B & B, contracting, subsidies or a salary earned off the farm by the wife, to keep them afloat financially.

Financial success in smallholding has little relationship to the area of land available; rather it is the vision, flair and determination of the smallholder that makes the difference. To generate a living wage from a small acreage is possible but extremely difficult, and only a few exceptional people have the determination, commitment and business acumen to achieve this.

Once accustomed to "earning a living", it can difficult to re-adjust to life on a much lower income, but this lies at the heart of the matter. Our personal view is that smallholding is not about maximising your income; rather it is about reducing the income that you actually need down to a manageable level. You can live, and live well, on a much lower income than before, once the smallholding itself is meeting many of the family's needs. Turn the question around, and ask instead "how much income do you really need?" For most people, the answer is that you do still need to earn some outside income, but nowhere near as much as the landless city dweller.

It's useful to think of smallholding in terms of a sliding scale of possibilities. At one extreme lies the search for complete self-sufficiency with independence from earned income; at the opposite extreme lies the endeavour to generate a living wage from a smallholding business. In reality, most people will position themselves partway along such a scale between the two extremes, whereby the activities and produce of the smallholding partly offset the cost of living there. To meet all their financial needs, the majority of smallholders do require additional income from some outside source such as part time or freelance work, grants and subsidies, or investment income from pension or redundancy.

A useful rule of thumb that we came across before our own move into smallholding is that, however self-sufficient we became, we should still expect the need to earn between one third and one half of our former income to remain financially viable. This turned out to be accurate in our case.

Let's now examine some of the ways in which the smallholder can save money.

Taxation

Paying income tax is universally unpopular, but you can reduce your liability, defer it, or even avoid paying any at all, by running your smallholding as an agricultural business. When you first register as a self-employed "farmer", you are allowed a period of five years to reach a profitable trading position. During this period, capital expenditure will be high as you invest in fencing, gates, buildings, livestock and machinery, which of course is the reason for this concession, and much of this expenditure is wholly or partly claimable as a business expense.

In the fifth year of trading you need to declare a profit, otherwise the tax inspector will no longer accept the enterprise as a viable business and will treat it as a hobby instead. By this stage you can be up and running with an income stream that meets this requirement, and provided the profit stays within your annual tax-free personal allowance, there's still no income tax to pay.

The advantage of running the smallholding as a business instead of a hobby is that a broad range of expenses can be offset as tax-allowable against income, including a proportion of the everyday household budget such as the running costs of the farmhouse provided the business office is located within it. Self-assessment enables the lay person to keep their own accounts and submit an annual return to the authorities. Rules and thresholds change but the principle holds that the smallholder can easily reduce or even avoid his liability to pay income tax.

Few smallholding businesses will reach the level of turnover at which registration for VAT becomes compulsory. However, you can choose to register for VAT irrespective of turnover, so that tax charged on expenses can be offset against the tax charged on sales. Many smallholding sales are zero rated as food products, so the practical effect is that your business can reclaim the VAT paid on expenses – a substantial saving. Against this must be set the burden of paperwork and legal responsibilities involved. It's your choice.

Excise duty is another tax that the smallholder can legally avoid paying by the simple expedient of home brewing. Brewing your own beer or country wine saves a considerable sum of money over the year compared to buying the commercial product. Quite apart from the profit required by brewer, publican, supermarket or other middlemen, there is the taxation burden which home brew completely avoids.

Above: A sideline business
Opposite left: A roadside board
Opposite right: Elderflower champagne is easy to make

Ageing estate car as workhorse vehicle

Depreciation

Depreciation is a factor that is often overlooked, particularly in respect of motor vehicles. Public transport tends to be limited or non-existent in rural areas, leaving most smallholders with little choice but to run a private car and/or utility vehicle. Part of the running costs and depreciation of these are claimable as a business expense.

The major cost of owning a vehicle from new, or nearly new, is depreciation. Buy a shiny new four-wheel-drive vehicle for say £30,000, sell it after four years and it will have depreciated to perhaps £15,000, a cost of more than £10 per day regardless of use. Fuel economy, maintenance and other costs are often secondary by comparison to such erosion of capital. For the smallholder operating on a tight budget, such expenditure makes no financial sense.

Depreciation strikes hardest in the early years, so for the lowest overall running costs, it makes sense to consider older vehicles of proven long-term quality. For example, for the last 27 years a succession of three ageing Volvo estate cars have served as our smallholding workhorse, each purchased when around 12 to 14 years old and eventually scrapped around 10 years later. The combined purchase cost of all three has been £2,800 so the depreciation works out at £104 a year so far, or just 28 pence a day - and the third one is still going strong!

Barter

Barter is a traditional means of exchanging goods and services in the countryside, and often involves something produced on the smallholding. As examples, we have bartered ducklings in exchange for the loan of a ram; two lambs plus some hay in exchange for a computer; a freezer lamb in exchange for a greenhouse; and helped a friend fix his roof timbers into position in return for his help spreading our delivery of ready-mixed concrete. There is plenty of scope for barter in country living, and smallholders should be alert for the opportunity.

The spirit of smallholding

Generally, when the focus is on maximising income from smallholding activities, there is a risk that the development of products and markets turns the smallholding into a hard-nosed business, which for us runs contrary to the spirit of the whole enterprise, and may defeat the point of moving to the country in the first place. Not everyone would agree with this, of course, but our approach can be summarised as follows:

- Focus on avoiding expenditure rather than generating income
- Avoid a mortgage if at all possible
- Legally avoid some taxes
- Minimise capital depreciation: buy quality second-hand vehicles/machinery
- Barter where possible
- Earn enough outside income to cover reduced living costs

Livestock

How much profit can be made from livestock such as cattle, goats, pigs, ducks, chickens, turkeys and so on? If we examine each of these in turn, we end up at much the same conclusion – that the careful smallholder can earn a modest profit from each, but certainly not enough to provide for the financial needs of the household.

This is the harsh reality that drives commercial farmers to keep ever-larger numbers of livestock in the search for economies of scale. However, once you begin to tread that route, finding enough customers for your premium price demands far more time and effort. That's not to say it can't be done; it can, but then you're sliding further along that scale towards running a business, instead of self-sufficiency.

Suppose that you intend to keep a small flock of 12 breeding ewes, aiming to produce around 20 fat lambs for private sale each year. The economics look something like this at the time of writing:

Above: Young lambs playing
Opposite: A smallholding business

Income:

20 lambs sold privately as freezer meat @ £90	£1,800

Costs:

3 lambs (out of the 20) for home consumption @90	£270
Hire of ram	£30
72 x bales hay @ £2.50	£180
30 x bales straw @ £2	£60
24 x bags organic feed @ £12	£288
20 x slaughter and butcher @ £16	£320
Veterinary costs, say	£100
Total costs	£1,248
Gross profit	£422

For simplicity, these figures ignore the minor costs and income of cull ewes and fleeces. You will also need to buy original stock and replacement breeding ewes (or grow your own); provide transport to and from the slaughterhouse and distribute to

Buying in weaners may be cheaper than breeding them

customers; collect and store feed and bedding; and provide for ongoing capital expenditure on fencing, gates, handling pens and lambing facilities. Last but not least, you need to find 17 customers for your lamb at a premium price in return for knowing where it comes from and how it has been reared.

There are potential savings to be made by harvesting your own hay and bedding, from your own resources or by using contractors, but clearly, sheep keeping on a small scale is not going to finance a life of idle luxury. A commercial rule-of-thumb is to earn £15 profit per ewe per year.

If sheep aren't going to make your fortune, what about pigs? Suppose that you buy a few weaners in the spring and fatten them for slaughter in the autumn. The economics look something like this at the time of writing for each pork pig of 150 pounds live weight:

Costs:		
Cost of weaner		£35
6 bags x organic feed @ £12		£72
Kill and butcher		£36
Bedding, sundries		£7
	Total	£150

In return for this you should receive around 90 pounds of pork in various forms.

For a bacon pig of 200 pounds live weight the economics look like this:

Costs:	
Cost of weaner	£35
9 bags x organic feed @£12	£108
Kill, butcher and cure	£100
Bedding, sundries	£7
Total	£250

In return for this you should receive around 120 pounds of bacon and ham. There are potential savings to be made by home slaughtering and curing if the produce is solely for your own consumption.

Might it be cheaper to keep a breeding sow or two, and produce your own weaners? We did this for a decade and found that overall it was marginally cheaper to buy in than to breed, but more significantly it saved a great deal of work. That's not to deny the pleasure and independence of breeding your own, but here we are focusing on financial aspects. If you keep two breeding sows – the minimum number, for a pig must not be kept alone - and each sow rears two litters of 10 piglets a year, you have to process and sell 40 pigs a year. That means a great deal of work, and a lot of meat to find customers for, so you risk becoming a pig farmer when all you really wanted was your own pork.

Typically we buy weaners in late April, and supplement their bought-in rations through the summer with fruit and vegetable waste from the garden, plus spent grains from home brewing of beer; then glut them with fallen apples and pears from our orchard through the autumn. In September some are killed and sold as pork, leaving the smallest and largest of the bunch to grow on. In October these last two are killed for our own use, the smaller for pork, the larger for bacon and ham. The result is an ample supply of joints, chops, cubes, mince, sausages, bacon, hams and gammon steaks to see us comfortably through the next 12 months, while the overall cost to us, balancing the income from those sold against expenditure, is close to zero – plus we have the muck to put back on the garden.

If we expand the numbers reared, the benefit of our home-produced feedstuffs would be diluted so that feed costs per pig would increase. You have only to look at the commercial pig sector to see the very narrow profit margins involved once all feedstuffs are bought in. The smallholder scores here precisely because a few pigs can be fed relatively cheaply on the waste products from other smallholding activities, but as numbers rise, this advantage quickly disappears in a reversal of the usual "economy of scale" thinking.

And this is where we came in!

Chapter 13

The self-sufficient kitchen

On a productive smallholding, it is possible to approach self-reliance in the kitchen all year round, producing high quality organic food at a fraction of the cost of buying ready made.

During a good growing season there will be more produce than you can possibly eat fresh, but there are few vegetables and fruits that cannot be preserved in some way. We prefer to eat vegetables in season and gathered fresh just before cooking, so that as the year passes we look forward to a particular crop and really enjoy it for the time it is available. However, it's good to have at least some vegetables stored as back-up, particularly during the "hungry gap" of late winter and early spring.

Storing vegetables

Most root vegetables can be stored by cellaring. Potatoes are easily stored in paper sacks kept in a cool but frost-free place. Most other roots dry out and shrivel if stored like this, so instead are buried in slightly damp sand or fine soil within tea chests or similar stout containers. Here in the mild south west of the UK, we mostly leave carrots, parsnips, turnips and celeriac in the ground where they have grown and lift as required; but of course the further north your garden, the higher the risk of frost damage.

Ball-headed cabbages can be uprooted and hung in a cool outbuilding to store for several weeks, even months if the heads are really solid. Onions are dried in the sun before plaiting into strings, then hung in a cool, airy place where they will keep until spring.

Peas and broad beans freeze well if picked young and tender, then podded and frozen straight away; tasting almost as good as fresh picked when cooked. It's often advised to blanche vegetables before freezing (plunge for a short time in boiling water) but this increases the workload and, being wet, they all stick together in a solid block. We find that without blanching, peas and beans freeze individually so you can take out the required amount from a bulk bag. French beans we also pick young and chop before freezing. Harvest regularly so that beans are never past their best. We don't find runner beans freeze well, so instead use the surplus as a main ingredient for pickles and chutneys.

Opposite: Carrots can either be cellared or left in the ground

Red, white and black currant harvest

Storing fruit

By planting a range of apple varieties, you can enjoy your own apples from late July through to March or April. We freeze some cookers, peeling and chopping them, then lightly cooking (to avoid browning) before freezing in plastic boxes. Knock the frozen blocks out to stack neatly in bulk bags, each block being enough to make an apple crumble or similar dessert. Pears are just peeled and cored before freezing in quarters in portion-sized bags, or on open trays before bagging together. Plums bottle well in Kilner jars preserved in a syrup, with added brandy if you like to make a special treat later in the year.

Of the soft fruits, black and red currents are rather fiddly to prepare for freezing by removing the stalks, but well worth the effort. Similarly, gooseberries need "topping and tailing". Raspberries just need picking, and our autumn varieties yield enough to enjoy straight off the bush while the surplus is frozen in portion-sized bags to enjoy throughout the year. Blackberries grow profusely on the smallholding and we gather plenty to freeze in bulk bags.

Strawberries are best enjoyed fresh, but can be preserved in ice-cream. Just whip up double and single cream together, mix in some strawberries and enough sugar to taste. How much cream will depend on how many strawberries there are – just keep tasting

Blackberries from the hedgerow

to see when it's right! Place the mixing bowl in the freezer and every half-hour or so as it hardens, stir the mixture round. Before it turns solid, place the ice cream in portion-size containers. When making gooseberry, black or red current ice cream, cook the fruit with sugar to taste for a short time before mixing with the whipped cream.

All fruits can also be preserved as wine.

Preserves

A whole range of preserves can be made with surplus garden produce. Strawberry jam is a must for Devon cream teas in the summer with scones and clotted cream. Add a few gooseberries to help the jam set (and to make it go further!). Black currants are high in pectin so their jam sets easily. Plum or damson jam is good too, but remove the stones before cooking.

A glut of cooking apples can be used to make a range of jellies to serve with sausages, cold meats, Sunday roasts, ice cream and Christmas dinner. Once you have put the best apples to store, peeled and chopped the large ones to freeze for puddings, and pressed more of them for juice or cider, you are left with the smaller, damaged or misshapen fruits to make jellies. Wash them, chop roughly, cover with water and cook until soft before straining through muslin. Jelly made with this juice will be simply apple jelly. For mint jelly, add fresh mint leaves to the juice and bring it to the boil. Remove from heat and stand to infuse. Remove the leaves, measure the liquid and add one pound of sugar to every pint of juice. Stir to dissolve the sugar then return to the boil and simmer until setting point is reached. Use different herbs for more jellies – rosemary, lemon balm or lemon geranium.

Pickle and chutney

Pickles or chutneys offer a delicious way to preserve vegetables, complementing cheese or ham with bread for a ploughman's lunch. There are many recipes for chutney but the basics are:

- five pounds of vegetables (chopped runner beans, onions, tomatoes etc.) or apples

- three-quarters-of-a-pint of vinegar

- six ounces of sugar

- optional spices: one tablespoon of dry mustard; two tablespoons of turmeric (dyes chutney yellow) or celery seed; four tablespoons of mustard seed - or chillies to taste! Place the whole spices in a square of muslin and remove after cooking.

Mix ingredients together and slowly bring to the boil. Simmer uncovered to reduce liquid for about 30 minutes before bottling.

Vegetables often benefit from brining before pickling, to reduce their water content and preserve a crunchy texture. Beetroot is an exception and these are boiled in water before slicing and pickling in spiced vinegar.

Use dry brining for vegetables with a high water content like marrow or cucumber. Layer the chopped vegetables with one or two ounces of salt to one pound of vegetables. Cover and leave for 24 hours. Drain and rinse well.

Use wet brining for firm vegetables like cauliflower and shallots. Cover the chopped vegetables or whole shallots with a brine solution of two ounces of salt to a pint of water for each pound of vegetables. A plate on top of the bowl keeps the vegetables submerged. Drain after 24 hours and rinse well.

The vinegar for pickling is usually spiced first. You can buy ready spiced vinegar but it's easy to make your own. Place two pints of vinegar in a saucepan then add one tablespoon each of cloves, mace, allspice and peppercorns, plus 2 bay leaves and a piece of cinnamon stick. Chilli and mustard seeds make a hotter mix if desired. Bring to the boil then leave overnight to infuse.

Pack the vegetables into the jars. Heat the spiced vinegar then pour into jars to fill the remaining space. Place the lids on and screw down tightly.

Home made jams, jellies, pickles and chutneys

Bread

Bread is something most people eat every day, "the staff of life", so it needs to be good and wholesome. It doesn't take much time to bake your own bread and it combines well with a day at home. The following quantities are only a guide; flour varies so be adaptable. The "strong" is important as this indicates a flour high in gluten that will rise well. The texture of the dough is key.

A basic loaf recipe:

- three pounds of organic strong white or wholemeal flour.

- two ounces of fresh yeast (or one ounce dried)

- one teaspoon of sugar

- one-and-a-half pints of lukewarm water

- four teaspoons of salt

Above left: Adding water plus yeast
Middle left: Turning out dough
Below left: Setting dough to rise

Above right: Mixing dough
Middle right: Kneading dough
Below right: Fresh baked bread

Granary style organic wholemeal loaf

Mix yeast with sugar and about half-a-pint of lukewarm water. Leave until frothing. Add salt to the dry flour and mix well together. Add frothy yeast to flour with most of the water and mix thoroughly. Add a little water at a time, until the dough is neither sticky or dry. Knead thoroughly until the dough feels smooth and "plastic".

Put to rise until double in size, knock back, place in tins, rise again before baking - electric oven, use 425-450 °F; gas oven, use mark 7 or 8. Bake until the loaf sounds hollow when tapped underneath.

Variations:
- use milk instead of water to give a softer bread
- use milk and egg/s giving an enriched dough for a Brioche loaf or Chelsea buns
- add savoury ingredients such as cheese or herbs
- add sweet ingredients such as dried fruit, sugar and spices
- shape the bread in many ways
- sprinkle surface with seeds before baking

Bottled beer stores well

Beer

Why not use beer kits? These are usually based on a tin of hopped malt extract, the processed ingredient that saves the bother of brewing. This convenience costs you dearly both in money and quality. They also rely heavily on added sugar to boost alcohol levels, contributing nothing to the body or flavour of the finished beer. Beer made from kits just cannot match the quality of properly brewed real ales.

Brewing your own beer from wholesome ingredients saves money compared to buying the commercial product. Apart from the profit taken by the brewer and middlemen, there is the taxation burden which home brew completely avoids. At the time of writing, top quality real ale can be brewed at home from raw ingredients costing about 32 pence a pint.

The raw ingredients of beer are malted barley, hops, yeast and water.

Pale malt is barley modified by malting, in which the grain is subjected to controlled levels of humidity and temperature for germination and growth to begin. The process is halted at the right stage by heating and drying, before the grains are crushed for efficient brewing.

Hops are the flower of a climbing vine. They are available in tightly compressed, vacuum packed blocks that store well and can be relied upon for quality. There are many varieties but Goldings and Fuggles are traditional and widely used.

Yeast is required to ferment dissolved sugars into alcohol. Dried beer yeast is convenient, reliable and gives first-class results.

Many other ingredients can be used to modify the taste, colour or strength of the finished beer. Crystal malt is heated at the end of the malting process to a higher temperature than pale malt, giving a darker, caramelised appearance. It is often added in small amounts to give a darker colour and sweeter flavour to the beer.

Crystal malt (left) and pale malt (right)

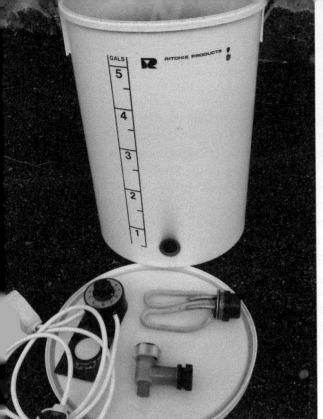

Above left: Bruheat boiler
Above right: Starting the mash

For mashing (see page 148) and subsequent boiling I use a purpose-made Bruheat boiler. This is a five gallon plastic bucket with an electric kettle element set into the base, controlled by a sensitive thermostat. A filter on the outlet tap prevents it blocking, home made from a short length of pipe or small plastic container.

It is possible to manage with a large preserving pan on the hob of a cooker, simply regulating the heat while stirring regularly to hold the temperature at the desired level. Obviously this requires more care and attention, but for that first attempt, why not?

Other essentials include an accurate thermometer, a siphon tube and two five gallon brewing bins. Do not use anything other than food grade brewing bins sold for the purpose - other plastics can contain poisonous compounds which may leach out into hot liquids.

The sugary liquids that result from brewing are an ideal medium for spoilage organisms to grow in; so make certain that all utensils and containers are both clean and sterile before use.

Raw ingredients for brewing

This basic recipe yields four gallons of delicious medium-strength traditional English bitter.

• Heat three-and-a-half gallons of water to 70 °C. Stir in six pounds of crushed pale plus eight ounces of crushed crystal malt, mixing well so that all the grain is wetted. Ensure the temperature has stabilised at 66/67 °C and maintain this level for one hour. This stage is known as mashing.

• Drain off the liquid, stir in a further two gallons of nearly boiling water and leave to soak for ten minutes. Drain off this second amount, adding to the first for a total of around four-and-a-half gallons of liquid, known as the wort.

• Add two-and-a-half ounces of Goldings hops to the wort and heat to a vigorous, rolling boil. Hold the boil for one hour.

• Drain off the wort, taking due care when handling scalding liquid. Cool to 25 °C, aerate by pouring splashily into another vessel, top up with tap water if necessary to four gallons and pitch (add) the yeast. Cover and stand in a warmish place (around 20 °C). Ferment until the frothy yeast head subsides after two days or so. Transfer to another vessel under airlock and continue fermentation for a further five days.

• Rack (siphon) into barrel or bottles, add finings (one dessert spoon of gelatin dissolved in a little very hot water) and half an ounce per gallon of priming sugar.

• The hardest part: try to leave for two or three weeks before drinking. Cheers!

Opposite top left: Mash temperature
Opposite top right: Boiling with hops
Opposite below left: Draining the wort
Opposite below right: Adding beer yeast

Some ingredients for country wine: lemon balm, strawberries, elderflower, peas, and rose petals

Country wines

Making country wine is so easy that anyone can do it. The main ingredients can be gathered from the smallholding, while the results are delicious and tax-free - so what are you waiting for?

Grapes contain everything necessary to make good wine in the right proportions. Most other fruits also yield good wine, but need added sugar for sufficient alcohol content. Vegetables provide flavour and aroma but contain less sugar, juice and acid than fruit, so require more additions to yield a balanced wine. Flowers and herbs provide flavour and aroma but little else, so all the sugars and acids required for a balanced wine must be added.

I suggest that you start by making fruit wines, moving on to other wines once some experience has been gained.

To make wine from anything, the flavours and sugars have to be extracted and modified into a sweet liquid for fermentation. Pressed fruit juice or cold soaked extract can be left to ferment through wild yeasts, although this runs a small risk that spoilage organisms may proliferate instead. The hot soak method uses boiling water poured over the ingredients to extract more efficiently than cold, whilst automatically destroying any spoilage organisms, but heat may modify the flavour of the resulting wine. Cooking is a method of last resort, only used for ingredients that are stubborn to yield an extract from less intrusive options.

Left: Straining off the juice   *Right: Cover and leave to stand*

All country wines need the addition of some sugar. I prefer to use a less refined organic sugar, in keeping with the natural qualities of country ingredients.

Most fruits and some flowers have natural yeasts associated with them, but these are destroyed by the extraction processes of hot soaking or cooking, so it then becomes necessary to add a dried wine yeast starter.

Pectin is the substance that makes jams set, and is present in all fruits and most vegetables. When released by heat, it creates an opaque haze that will not clear from the finished wine. To avoid this, add a teaspoon of pectin-destroying enzyme after using hot soak or cooking methods of extraction.

As for beer making, all utensils and vessels should be clean and sterile. Be methodical, clean everything after use, sterilise before use, and you should never have to pour any of your efforts down the drain!

Here's the basic method for redcurrant wine, with flavour extracted by the hot soak method. This makes a tangy, rich red wine that can be superb in a good year:

- three to four pounds of red currants

- two-and-three-quarter pounds of sugar

- pectin-destroying enzyme

- wine yeast

152

Crush the fruit thoroughly with a potato masher and pour on six pints of boiling water. Cover and leave to soak for two days, stirring occasionally. Strain off the liquid, squeezing as much from the pulp as possible, and top up to one gallon using cooled boiled water. Add the sugar and stir well to dissolve. Add pectin-destroying enzyme plus wine yeast, cover and leave to ferment in a warmish place (around 20 °C). Use a purpose-made plastic brewing bin with a snap-on lid. After a week or so, the vigorous fermentation subsides; now stir gently, pour into a gallon demijohn and fit an airlock.

When activity reduces to an occasional bubble after several more weeks, sample, and if too dry for your taste, dissolve a little more sugar and leave to continue the fermentation. Repeat until you are happy with the result. Once all activity has ceased, carefully siphon off the liquid into a fresh demijohn leaving only the sediment behind. Top up this new jar to the neck with cooled boiled water and re-fit the airlock. Leave to mature before bottling at around six months old.

Sugar levels can be adjusted as follows:

- two-and-a-half pounds of sugar per gallon gives a dry wine

- three pounds of sugar gives a medium wine

- three-and-a-half pounds of sugar gives a sweet wine

These are total amounts and allowance must be made for sugar extracted from the main ingredients: perhaps quarter to half a pound from ripe fruits. A wine that is too dry can easily be sweetened by adding sugar to taste as just described, whereas an over-sweet wine cannot be made any drier.

Chapter 14

The Flow

Among many books that we read before our own move into smallholding was *"Living on a little land"* by Patrick Rivers, first published in 1978. It's the honest story of a middle-aged couple pushed by redundancy away from the city to a derelict cottage in the country, where they set about taming the neglected land in search of self sufficiency on a low income. Within it's pages we found the following paragraphs:

> *Early in our first spring here I reached a point close to despair, for we did not seem equal to the task we had set ourselves. Then at the critical moment, two strong American students came into our lives, stayed first a day, then a week, then two weeks, choosing to help us rather than continue their walking holiday. Together they accomplished the impossible: they shifted mountains of scrap, dug ground and cleared scrub. When they left, they did so grateful and refreshed, and - so they told us - with a new outlook on work, and on the direction of their lives.*
>
> *From then on, help flowed our way. When our new solar heating refused to work, a practical physicist materialised and remade the system; when a pile of huge rocks fell on our vegetable terraces, five hefty men appeared and heaved them away. The very week we were ready for our Jersey calf a top breeder telephoned, offering us one which we promptly bought. When our income has run dry, work has appeared; and time and again, when our spirits have sunk, someone has arrived, telephoned or written to inspire us.*
>
> *The Flow is very important if you are to succeed in living on a little land. For this much we have discovered: that if you are doing what feels right - right for you and right for something greater than yourself - there is a strong possibility that help will flow to you, and that coincidences will punctuate your life, along with other happenings, which defy rational explanation. You have to expect the inexplicable to happen, and yet not wait for it passively - rather, you must go forth to meet it, working for it, recognising opportunity and taking chances."*

At that time I just smiled and turned the page, unable to take it seriously.

But I was wrong.

We had scarcely moved into our new home when an elderly man called Bill Finnemore knocked at the door, ostensibly to sell us firewood, but really to enquire about renting some of our grazing. Discreet enquiries revealed that he had recently retired from farming to a bungalow in the nearby village. Having two daughters but no son to carry on the farm, he had sold everything except a small flock of sheep which he kept to pass the time. We were keen to learn about sheep and here was a man with a lifetime's knowledge of them. Participating in the annual round of sheep management

Learning from Bill, an experienced shepherd

with his lambing flock, before we started to acquire our own, was priceless experience and something we could never have planned to happen - it just did.

The previous occupier of our home was a tenant, an elderly man who enquired if we knew how to look after the ancient grapevine in the lean-to greenhouse. We didn't, so he showed us how to scrape off the loose bark each winter, and prune back to just two or three buds at the base of each lateral shoot. In due course he vacated and we moved in during November.

It was a sunny day in March when we first realised that the drain from the toilet was blocked. The drainage arrangements had euphemistically been described as "private" by the estate agent, and there was presumed to be a septic tank at an unknown location in the field. We tried in vain to rod the blockage clear, and observed that if we could only speak to the former tenant, he might know the layout. Scarcely had these words been exchanged when his old car turned into the driveway and halted beside us - he'd called to check that we'd pruned the vine!

And yes, he knew exactly where the drains ran, as he'd often needed to clear them himself. We will spare you the details, but suffice it to say that "inadequate sanitation" later qualified us for a substantial local authority grant towards the renovation of our home, at a time when these were virtually unobtainable on any other grounds.

One of the items on our shopping list when searching for property was the potential for a sizeable pond. On the smallholding that we bought, the obvious choice was a rather more ambitious project. Floodwater would often form a temporary lake of

Left: the greenhouse grapevine   *Right: The new lake*

up to one and a half acres in a riverside meadow, and we saw the opportunity to convert this into a permanent water feature. The available grants were researched but none were appropriate. Undaunted, we decided to go ahead anyway, obtained planning permission and booked a local contractor. Work was about to start when a local newspaper reported the launch of a brand new scheme to fund environmental and landscape improvements. We applied and were quickly offered 50% of the cost, even as the diggers carved out the new lake.

I could go on to describe many other examples of The Flow at work. We were slow to recognise it at first, but as the evidence accumulated, eventually we had to reach for that book again and re-read the section with more open minds. The name passed into our vocabulary, so that when the unexpected happens to support what we are trying to achieve, we explain it to each other by saying simply "it's The Flow".

Now, you may be thinking that this is all down to coincidence, luck, fate, serendipity, or whatever other word you choose to describe the surprises that life occasionally bestows. For that to be true, such chance happenings would occur at random and irrespective of your own efforts - but they don't. If you sit in a chair and complain, nothing positive comes your way. The Flow doesn't seem to work if you wait for it passively; it only works when you are actively pursuing something, putting time and energy into it, reaching out to others, seeking opportunities and advice.

I don't mean to suggest there is anything mystical or quasi-religious about this, although it might be described as the spiritual side of smallholding. There is strong evidence in the field of psychology that a positive, outward looking projection of ideas

157

encourages a more constructive and helpful response from others, a feed-back effect that reinforces the advantages of a positive approach. Perhaps The Flow results from the broader projection of this effect, spreading beyond those with whom you are in personal contact to reach and affect a wider audience.

Whatever the explanation, we do now believe that when you seek to achieve something that is, in the words of Patrick Rivers "right for you and right for something greater than yourself", help will often flow to you in a way that defies rational explanation. So if smallholding is right for you, and if you approach it in a positive and constructive fashion that respects the natural world and your place within it, be prepared to expect the unexpected.

Or to put it another way:

"may The Flow be with you"!

Resources

"A Start in Smallholding" by Alan Beat, published by Smallholding Press.
ISBN 0-9546923-0-6
The story of the author's move into smallholding.
Available through: www.thebridgemill.org.uk
or Amazon: www.amazon.co.uk

The Bridge Mill website: www.thebridgemill.org.uk
Details of open days and educational visits to the author's smallholding

World Wide Opportunities on Organic Farms www.wwoof.org.uk
Work experience - your labour in exchange for tuition plus board and lodging:
WWOOF, PO Box 2154, Winslow, Bucks MK18 3WS

HelpX: www.helpx.net
Similar to WWOOF above but Internet-based

Henry Doubleday Research Association www.gardenorganic.co.uk
Now using the working name of "Garden Organic", Ryton Organic Gardens,
Coventry, Warwickshire CV8 3LG Tel: 02476 303517

Soil Association www.soilassociation.org
Organic standards and and certification:
South Plaza, Marlborough Street, Bristol BS1 3NX
Tel: 0117 314 5000

Wholesome Food Association: www.wholesome-food.org
Alternative certification scheme

DEFRA www.defra.gov.uk
For rules, regulations, codes of practice, grants, and all official stuff: or locate
the regional office via your local telephone directory.

Country Smallholding magazine www.countrysmallholding.com
Archant South West, Fair Oak Close, Exeter Airport Business Park, Clyst Honiton,
Exeter EX5 2UL
Tel: 01392 447766

The Smallholder magazine www.smallholder.co.uk
3 Falmouth Business Park, Bickland Water Road, Falmouth, Cornwall TR11 4SZ
Tel: 01326 213340

Home Farmer magazine www.homefarmer.co.uk
Tel: 01772 633444

Lightning Source UK Ltd.
Milton Keynes UK
UKHW051129210420
362006UK00003B/65